工业遗产影像志

江城之虹

胡晨冉 摄

华中科技大学出版社
http://press.hust.edu.cn
中国·武汉

内容提要

本书为一部关于武汉工业遗产空间的摄影集，通过作者拍摄的几百张实景照片与部分历史档案图片，集中展示了新时代武汉工业遗产蝶变更新的全貌，呈现了武汉作为历史文化名城的形象。

本书涵盖了武汉著名的工业遗产，如汉阳造文化创意产业园（武汉鹦鹉磁带厂旧址）、四美塘铁路遗址文化公园、杨泗港都市 T 台、站前花街、宗关水厂等。

图书在版编目（CIP）数据

江城之虹：工业遗产影像志 / 胡晨冉摄 . -- 武汉：华中科技大学出版社，2025.7. -- ISBN 978-7-5772-1925-7

Ⅰ . TU27-64

中国国家版本馆 CIP 数据核字第 2025YD3929 号

江城之虹：工业遗产影像志   胡晨冉 摄
Jiangcheng zhi Hong:Gongye Yichan Yingxiangzhi

策划编辑：金　紫
责任编辑：梁　任
装帧设计：金　金
责任监印：朱　玢
出版发行：华中科技大学出版社（中国·武汉）　　电　　话：（027）81321913
　　　　　武汉市东湖新技术开发区华工科技园　　邮　　编：430223
录　　排：天津清格印象文化传播有限公司
印　　刷：湖北金港彩印有限公司
开　　本：889mm×1194mm　1/16
印　　张：13
字　　数：125 千字
版　　次：2025 年 7 月第 1 版第 1 次印刷
定　　价：168.00 元

创作缘起

我是新闻系的一名学生，是介于文字工作者与影像工作者之间的跨界者。

对工业遗产的关注，很大程度上是因为我的家乡宁波也是一座工业城市，而武汉作为我大学求学之地，其丰富的工业遗产更是激发了我许多创作上的灵感，这也是我想要为这座城市的工业遗产留下影像资料的原因。

我并非诗人，但在这本摄影集中却收录了十几首诗歌。严格意义上讲，这些诗歌其实并不能完全算是我个人的独立创作，而是我与 DeepSeek 共同完成的一场"试验"。我在与 DeepSeek 的对话过程中，首先描述了我所拍摄场景的样态，并创作了诗歌的开篇，而后向 DeepSeek 发布了我希望呈现出何种意象的指令，最终便生成了具有鲜明 DeepSeek 风格的诗句。

当然，DeepSeek 毕竟未能亲临现场，而我所认为的诗歌又必须具备一定的真实性和可读性，所以我会对 DeepSeek 生成的诗句进行事实性和艺术性的修改。正因如此，在我看来，最终呈现在本书里的诗篇，实际上更接近于依托"人机互动"而形成的"试验文本"。

我不知道这是不是人类历史上首部创作者与大语言模型合作完成的诗集，但这的的确确是我的一次"胆大包天"的艺术尝试和诗意冒险。作为初出茅庐的新人，我认为，工业化是人类现代文明的主旋律之一，而让象征中国工业文明源流的工业遗产与担负新型工业化使命的 DeepSeek 相遇，也许是我为面向未来的现代化之梦所能贡献的一点绵薄之力。

在中国文化史上，向来有"左图右史"的文学传统。我希望，这本摄影集里的图片和诗歌，可以作为一个共同呈现的艺术整体，在成为历史记录的同时，也能够穿梭于时空，通向更为遥远的未来。

胡晨冉

序

在光影中追寻城市的呼吸

胡晨冉同学邀我为她的摄影集《江城之虹：工业遗产影像志》作序时，我正伏案翻阅 20 世纪下半叶武汉几家老工厂微黑的蓝图档案，那些用鸭嘴笔绘制的蒸汽管道轮廓，恰与窗外杨泗港长江大桥的钢铁弧线形成某种时空对话。当转头凝视年轻人用数字镜头凝固的新生工业景观时，我忽然想起三四十年前那个骑着二八自行车、背着武汉生产的友谊牌相机穿梭在硚口码头、龟山脚下、青山江边的自己——原来这座城市的文化基因，始终在代际传承中焕发着蓬勃生机。

作为在武汉工业遗产保护领域耕耘半生的地方志与城市史工作者，我亲历过太多令人扼腕的消逝。我们曾为汉阳铁厂等多个工业遗产的保护、改造而呼吁、奔走，但由于时代的局限性，武汉大量的工业遗产仍不免湮灭于滚滚的历史车轮之下。正因如此，当我翻阅这本《江城之虹：工业遗产影像志》摄影集时，指尖竟有些微微发颤——这些影像构建的，不正是我们这代人梦寐以求的"另一种可能"吗？

胡晨冉同学镜头下的工业遗产，呈现出超越时空的叙事智慧。在青山区红房子改造的社区图书馆里，阳光穿过保留的混凝土花窗，这是穿越时空的无声对话；在汉阳造文化创意产业园，一只白猫在红墙的映衬下格外灵动，这些影像未流于简单的今昔对比，而是用光影编织出工业文明的传承密码，让观者看见钢水淬炼的火花如何转化为文化创新的热能。

年轻人独特的视觉语言令我耳目一新。在拍摄汉口平和打包厂改造的钻石艺术馆时，她选择在梅雨季节取景，水渍斑驳的墙面恰似当年棉花打包工额头的汗珠，而新时代的新工人——快递员正驾驶着一辆堆满货物的电动车疾驰而过。这种将历史质感与当代审美熔铸一炉的创作理念，让我想起法国摄影师尤金·阿杰特记录老巴黎的执着，却又带着数字时代特有的像素诗意。《江城之虹：工业遗产影像志》这部摄影集再次告诉我，文化遗产还可以玩得这样美！

这部摄影集最珍贵的是其展现了工业遗产作为"城市有机体"的生命律动。杨泗港都市 T 台的步道上，一位散步者的悠闲身影仿佛正告诉我们什么是"人民城市"；红 T 时尚创意街区里来来往往的市民则为这座城市的工业记忆注入新的灵魂。诚然，真正的遗产保护不是制作标本，而是培育能在当代社会存活的"文化干细胞"。胡晨冉同学用镜头证明，当工业记忆转化为市民的日常体验后，那些沉默的钢铁巨构才能真正获得新生。

作为看着武汉工业版图变迁的"老武汉人",我格外珍视影像中的人文温度。在胡晨冉同学拍摄的粤汉铁路火车轮渡码头旧址中,两辆停放在大堤上的自行车,无声地反映了废墟作为景观的"人气",这比任何文字都更能诠释遗产传承的真谛。又如在站前花街,游客或驻足欣赏,或热烈交谈,正是对当下人民群众美好生活的完美诠释。胡晨冉同学的视觉人类学眼光,使这部摄影集具有文献与艺术的双重价值。这种立体化的观察方式,正是做田野调查需要的多维视角。

翻阅这些作品时,我常想起林徽因当年为挽救北京古城墙的奔走疾呼。不同的是,新时代的遗产保护者拥有了更丰富的表达工具。胡晨冉同学采用的影像叙事手法,既避免了伤感的怀旧情绪,又跳出了冰冷的学术框架,展现出 Z 世代特有的文化自信。她在后记中写道:"这部摄影集是我的双城记,更是一封来自新时代的回信。"我认为,这是自觉的传承意识,是武汉工业精神得以永续的关键。此时此刻,让我想起我们文化自觉者经常念叨的艾青的诗句:"为什么我的眼里常含泪水?因为我对这土地爱得深沉 ……"不同的是,新时代青年的爱,已化作更具建设性的文化自觉。

作为武汉从"工业重镇"到"设计之都"蜕变的见证者,我深信胡晨冉同学这代人的视觉书写,正在重新定义城市记忆的载体。当工业遗产不再是教科书里的铅字与图纸,而成为可感知、可互动、可传承的立体记忆场时,武汉的"东方芝加哥"魂魄便获得了真正的永生。这或许就是"江城之虹"的深意——工业文明经过时代的风雨,终将在江城武汉焕发出属于未来的七彩光芒。

临窗远眺,长江上货轮的汽笛声与江汉关钟楼的悠扬旋律交织成歌,合上这部充满朝气的摄影集,我看见年轻的手正接过历史的镜头,继续书写这座城市的光影传奇。这何尝不是最动人的文化传承?年轻的胡晨冉同学,请继续用你的镜头说话,让世界看见武汉这座"英雄城市"钢铁般的柔情。

原武汉地方志办公室副巡视员、武汉市国家历史文化名城保护委员会顾问

王汗吾

乙巳仲春于汉口

目录

楚天 181 文化创意产业园

楚天 181 文化创意产业园位于武汉市武昌区东湖路 181 号，由湖北日报传媒集团倾力打造。园区前身为湖北日报传媒集团的印刷厂房，占地约 40,000 平方米。楚天 181 文化创意产业园于 2011 年 7 月正式开园，是全国首家以现代传媒为主体的特色文化创意产业园。

Chutian 181 Cultural and Creative Industrial Park

Chutian 181 Cultural and Creative Industrial Park is located at No. 181 Donghu Road, Wuchang District, Wuhan City, proudly presented by Hubei Daily Media Group. The park was originally the group's printing plant, covering approximately 40,000 square meters. Chutian 181 Cultural and Creative Industrial Park officially opened in July 2011, it is notable as the first cultural and creative industrial park in China focused primarily on modern media.

印刷厂即兴曲

铅印车间的噪声把暮色敲成铜箔时

旧齿轮正从红砖缝里析出褐色的糖

传送带改造成空中栈道

断齿的轴承托举着玻璃匣——

那里液态的霓虹正被注入石膏模特

咖啡渍在车间图纸上晕开

方程式长出青苔触角

工程师的算盘珠被穿成风铃

每粒木珠里都蜷缩着八十年代的汗碱

涂鸦颜料正沿着消防栓攀缘

覆盖那些钙化的生产标语

装订机切削下的纸屑

在展厅里悬浮成白色的蒲公英

穿堂风掀起亚麻窗帘的一角

露出钢架结构嫁接的星空

锈蚀的配电箱深处

总闸仍卡在一九九三年的某个下午

此刻旋转门切割着时间

把工业咏叹调拌入电子混音

当互联网成为媒介时

纸张不会消失

只是所有的寂静

都开始共振

The Printing House Improvisation

When the clamor of the lead printing workshop hammers twilight
Into copper foils,

Old gears precipitate brown sugar from the cracks of red bricks.

Conveyor belts transform into air stacks,

While broken–toothed bearings cradle glass cases—
Where liquid neon flows into plaster mannequins.

Coffee stains swirl on shop drawings,

And equations sprout mossy tentacles.

Engineers' abacus beads thread themselves into wind chimes,

Each wooden bead is curled up with the sweat from the 1980s.

Graffiti paint ascends along fire hydrants,

Covering calcified production slogans.

The scraps of paper,

Cut by the bindery machine,

Hang suspended like white dandelions in the gallery.

The wind in the hall lifts the corners of linen curtains,

Revealing a starry sky grafted onto a steel frame.

Deep within the rusted switchboard,

The main switch remains stuck in an afternoon in 1993.

The revolving door slices through time,

Blending industrial arias with electronic remixes.

As the internet emerges as a medium,

Paper won't vanish;

Rather,

All the silences begin to resonate.

粤汉铁路火车轮渡码头旧址

粤汉铁路火车轮渡码头旧址位于武昌江滩徐家棚江边，是粤汉铁路的重要遗存。粤汉铁路始建于1900年，竣工于1936年。1937年3月10日，京汉铁路与粤汉铁路通过徐家棚站码头与汉口的刘家庙站码头的火车轮渡首次连成一体，实现了中国南北交通大动脉的无缝连接。码头的栈桥和铁轨遗迹在长江进入枯水期时露出全貌，锈迹斑斑的铁轨和斑驳的水泥与钢架，见证了武汉的百年变迁。这些遗迹不仅是工业历史的见证，也是市民们探寻历史、缅怀过去的热门打卡地。其中，用于支撑钢架的立柱大部分为百年以上的老铁轨，约1/3是汉阳铁厂造的铁轨，更显珍贵。如今，这里已成为粤汉铁路遗址公园的一部分，吸引着众多市民和游客前来游览和拍照。

Yue-Han Railway Train Ferry Terminal Historic Site

The Yue-Han Railway Train Ferry Terminal Historic Site is located on the riverside of Xujiapeng at Wuchang Yangtze River Beach. This site represents an important remnant of the Yue-Han Railway, which was constructed between 1900 and 1936. On March 10, 1937, the Jing-Han Railway and Yue-Han Railway were first interconnected through a train ferry service between Xujiapeng Wharf Station and Liujiamiao Wharf Station in Hankou. This historic achievement seamlessly linked China's north-south transportation artery. During the Yangtze River's dry season, the weathered remains of the wharf's trestles and tracks emerge in their entirety-rusted rails, crumbling concrete, and corroded steel girders standing as testament to Wuhan's century of transformation. These relics not only bear witness to industrial history but have also become popular spots for citizens to explore the past and revisit their heritage.Notably, many of the supporting columns of the steel frame are constructed from century-old railroad tracks, with about one-third originating from the Hanyang Iron Works, which makes them especially valuable. Today, this historic site forms part of the Yue-Han Railway Heritage Park, attracting numerous citizens and tourists eager to engage to visit and take photos.

铁轨显影术

当潮水退成显影液
江滩便浮出钢的脊椎骨
锈蚀的摩斯密码
在钢轨上持续发酵

汉阳铁厂的钢在浪纹里返祖
每道氧化褶皱都被拓印
蒸汽锤击打长江的平仄
此刻百年钢梁正以铁锈速率
将渡轮汽笛翻译成碳酸钙结晶

游客们用快门收割磷光
光绘列车正穿透防波堤的暗房
被切割的钢轨斜插入江心
像半截未发送的密电
而淤泥深处，被遗忘的铆钉
仍在叩击地质层的摩斯密码

老船工把缆绳结系入花岗岩的纹路
江水便咬出新月形齿痕
当枯水季的栈桥在暗箱中缓慢显影
所有快门沉入江底打捞
皱褶的历史

铁轨的伤口渗出银盐
冲洗粤汉线褪色的底片——
钢质血管的走向
原来是长江在显影液里
反复成像的暗色脊梁

The Art of Railroad Track Visualization

When the tide recedes into a developing
Liquid,

The spine of steel emerges from the
Yangtze River Beach,

The rusted Morse code continues to
Ferment on the steel rails.

The steel of Hanyang Iron Works returns to
Its roots in the waves,

Each oxidized crease is imprinted.

By the steam hammer striking the Yangtze
River,

At this moment,

The century-old steel beams is translating
The ferry whistle into calcium carbonate
Crystals at the rate of rust.

Tourists harvest phosphorescence with
Their shutters,

While light-painted trains penetrate the
Darkroom of the breakwater.

The cut rails slant into the center of the
River,

Like half of an unsent telegram,

And deep within the silt,

Forgotten rivets still tap out the Morse code
Of geological layer.

When the old boatman ties the mooring ropes
Into the granite's creases,
The river's teeth bite out crescent shaped
Indentations.
As the low-water pier slowly develops in the
Camera obsura,
All the shutters sink to the riverbed fishing for
The crumpled history.

The wounds of the railway tracks seep out silver
Salt,
Washing the faded negatives of the Yue-Han
Line—
The direction of the steel veins,
Is the dark spine of the Yangtze River repeatedly
Imaged in the developing solution.

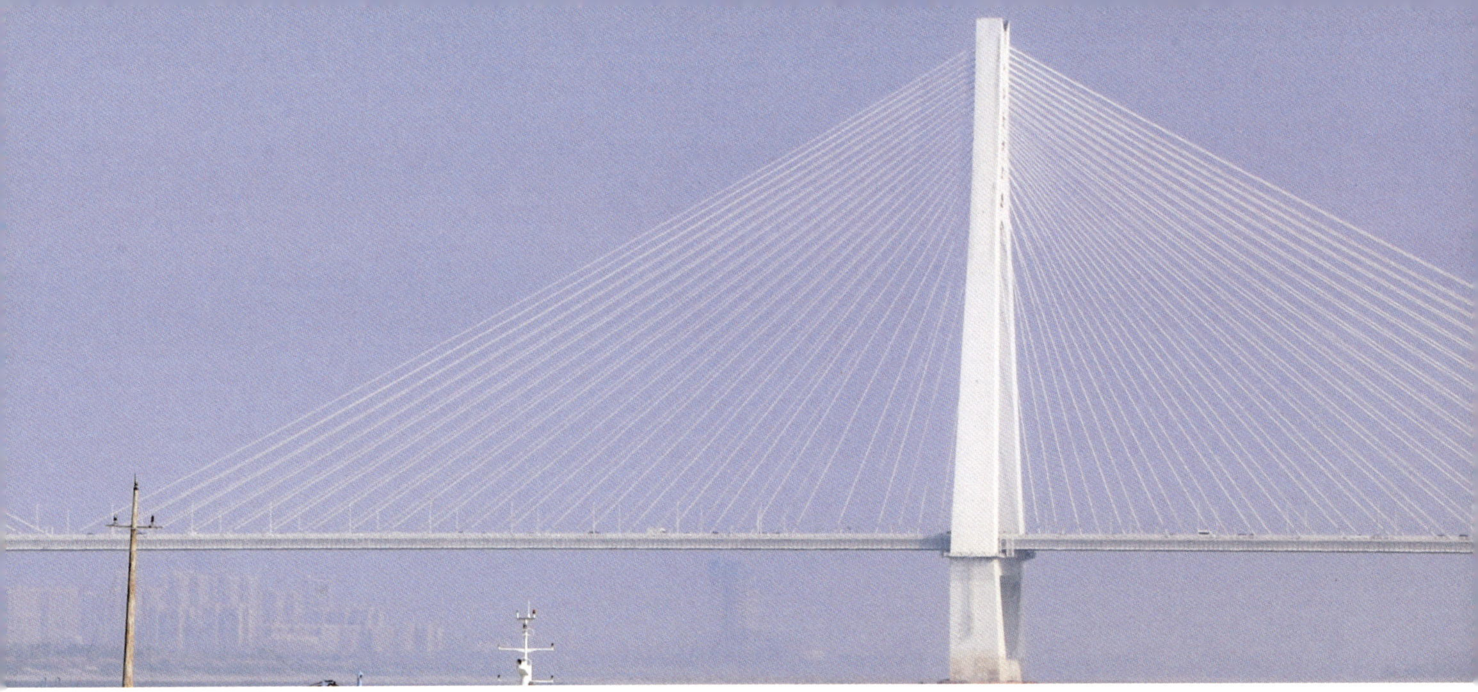

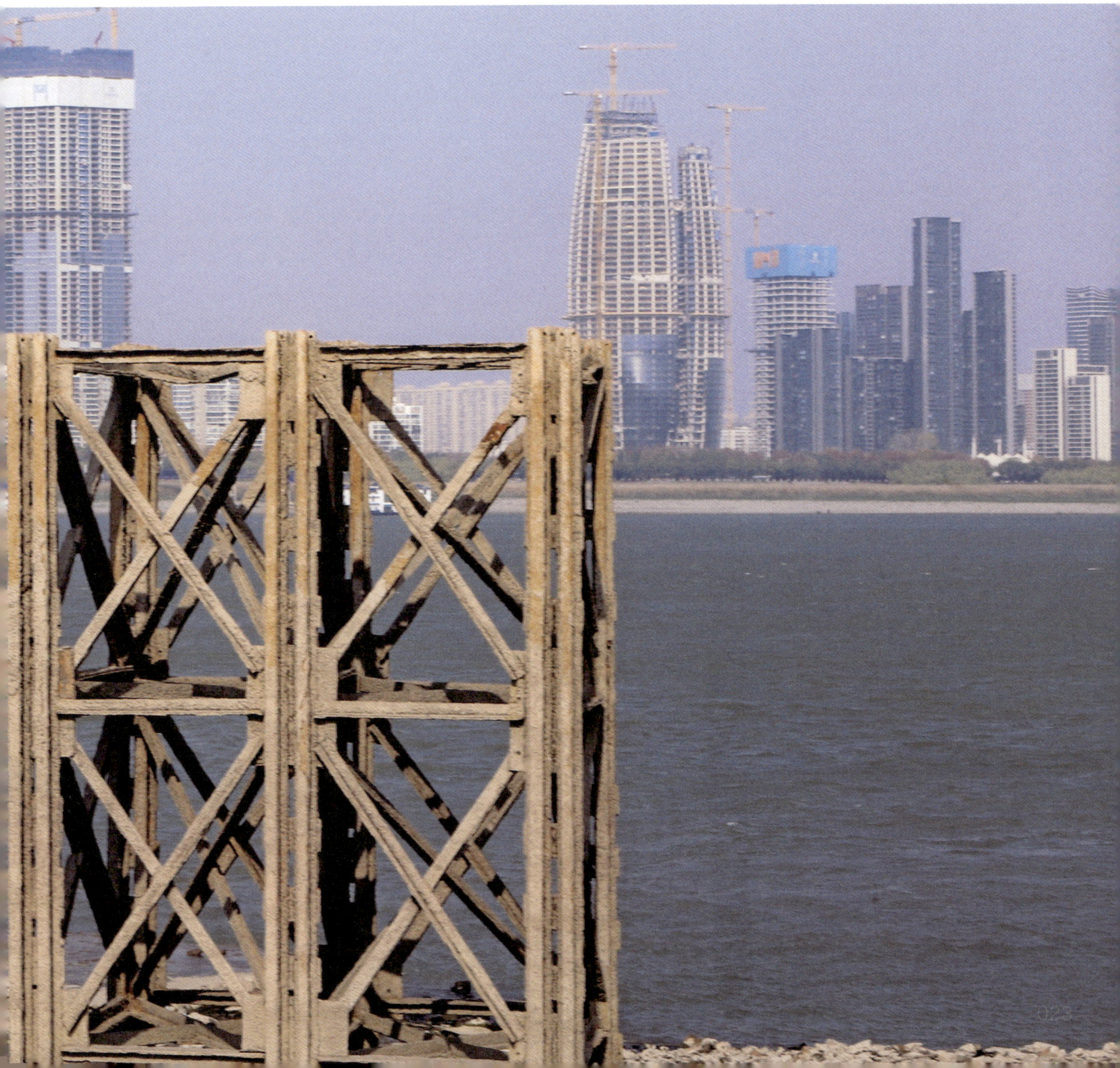

Big House 当代艺术中心

Big House 当代艺术中心位于武昌区临江大道 76 号。这里曾是 1915 年始建的武昌第一纱厂，见证了武汉纺织业的辉煌，其保存完好的建筑群更是堪称武汉近代工业遗产的代表作之一。如今，这座百年历史的办公楼被改造成为 Big House 当代艺术中心，成为武汉重要的公共文化空间。

Big House 当代艺术中心保留了典型的近代巴洛克风格建筑，造型精美且大气，具有鲜明的欧陆风格。园区内不仅有艺术展厅，还设有画廊、酒窖和西餐厅。这里不仅是艺术爱好者的天堂，也是市民和游客喜爱的"网红打卡地"。Big House 当代艺术中心以其独特的建筑风格和丰富的文化活动，成为武汉与世界交流的重要窗口。

Big House Contemporary Art Center

Big House Contemporary Art Center is located at No. 76 Linjiang Avenue in Wuchang District. It was once the Wuchang No. 1 Yarn Factory, established in 1915. It witnessed the glory of the textile industry in Wuhan, and the well-preserved architectural complex is regarded as one of the representative masterpieces of modern industrial heritage in Wuhan. Today, this historic building has been transformed into Big House Contemporary Art Center, an important public cultural space in Wuhan.

Big House Contemporary Art Center showcases typical modern Baroque architectures, characterized by exquisite and generous designs as well as a distinctive continental style. In addition to art exhibition halls, Big House Contemporary Art Center features galleries, wine cellars, and western restaurants, making it a heaven for art lovers and a popular destination for both citizens and tourists. With its unique architectural style and vibrant cultural activities, Big House Contemporary Art Center has become a crucial platform for Wuhan to connect with the world.

时间织就的茧

南洋风格的墙缝里
沉睡着一百年的灵魂
巴洛克式的钟楼伸入天空
像一位暮年绅士
仍倔强地维持着旧时代的体面

钢筋与花蔓在裂缝处亲吻
纺织机的轰鸣化作悬铃木的私语
黄昏漫过江景阳台
将一间间房屋熔成流动的琥珀

曾经裹挟棉絮起舞的气流
此刻正托起一幅未干的油画
月光在仓库深处织网
捕捉散落的钢琴键

空荡荡的穹顶悬着巨型艺术装置
线条垂落的姿态像极了当年的棉纱瀑布
消防楼梯盘旋成五线谱
铸铁通风管仍记得女工们哼唱的民谣旋律
某个被磨亮的黄铜旋钮上
指纹与指纹隔着世纪重叠

这里的时间是螺旋状的线轴
把工业文明的重量
纺成了轻盈的艺术纤维
在岁月长河的宏大叙事
与市井巷陌的烟火气息交织中
完成一场寂静的重生

The Cocoon of Time

In the cracks of the Nanyang style walls sleeps the soul of a
Hundred years.

The Baroque style clock tower stretches into the sky,

Like a gentleman in his twilight years,

Still stubbornly upholding the style of times past.

Steel bars and flower vines kiss at the fissures,

The roar of the textile machines transforms into the whispers of
The plane trees.

Dusk spreads over the river balcony,

Melting the houses into flowing amber.

The air currents that once danced with cotton now support a
Painting that has yet to dry.

Moonlight weaves a web deep within the warehouse,

Capturing scattered piano keys.

A giant art installation hangs from the empty vaulted ceiling,

Lines cascading like a waterfall of cotton yarn.

The fire stairs hover in a pentatonic pattern,

While cast-iron ventilation ducts still recall the melodies sung by
The women workers.

On polished brass knobs,

Fingerprints overlap with those from centuries past.

Here,

Time is a spiral spool,

Spinning the weight of industrial civilization into the light fibers of
Art.

In the dialogue between the great tales of the long river of time
And the bustling atmosphere of the city's streets and alleys,

The rebirth of silence finds its completion.

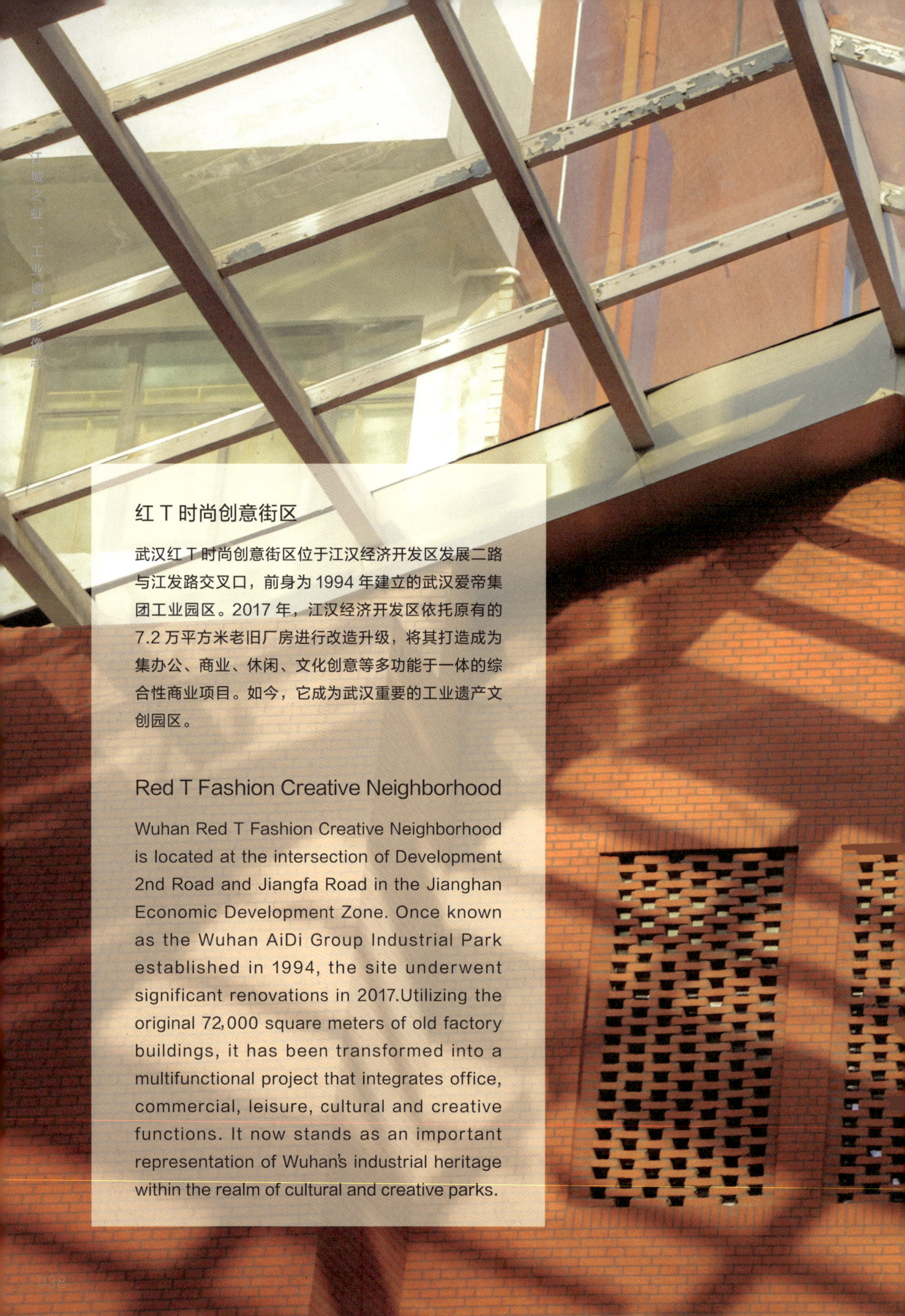

红 T 时尚创意街区

武汉红 T 时尚创意街区位于江汉经济开发区发展二路与江发路交叉口，前身为 1994 年建立的武汉爱帝集团工业园区。2017 年，江汉经济开发区依托原有的 7.2 万平方米老旧厂房进行改造升级，将其打造成为集办公、商业、休闲、文化创意等多功能于一体的综合性商业项目。如今，它成为武汉重要的工业遗产文创园区。

Red T Fashion Creative Neighborhood

Wuhan Red T Fashion Creative Neighborhood is located at the intersection of Development 2nd Road and Jiangfa Road in the Jianghan Economic Development Zone. Once known as the Wuhan AiDi Group Industrial Park established in 1994, the site underwent significant renovations in 2017.Utilizing the original 72,000 square meters of old factory buildings, it has been transformed into a multifunctional project that integrates office, commercial, leisure, cultural and creative functions. It now stands as an important representation of Wuhan's industrial heritage within the realm of cultural and creative parks.

光的等高线

正午的缝纫机把阳光扎成金线

穿过红砖厂房的针眼

那些曾被蒸汽熨烫的屋顶

正在为青春裁剪光的等高线

废弃的熨斗塔楼里

霓虹

旋转而成圆窗上的瓦砾

暖风推开锯齿状天窗

年轻的生命正用碳纤维

编织云影的经纬线

书香氤氲于混凝土立柱时

三五个剪影正踩着光瀑

把冬日的色温调至香槟模式

斜射光切开中庭的旋转楼梯

不锈钢扶手上流淌着

去年末拆封的时装秀背景音乐

涂鸦墙剥落处渗出铁锈

与胭脂在转角完成色彩反应

当晾衣绳改装的灯带亮起

悬吊的缝纫梭突然导通电流

所有阴影都褪去

唯有新栽的琴叶榕在备忘录里

记下此刻

暖阳正给老厂牌缝制春日的新装

The Contours of Light

At noon,

The sewing machines tie sunlight into golden threads,

Through the eye of a needle in a red-brick factory.

Roofs once ironed by steam now sculpt the contours of Light for youth.

Inside the abandoned ironing towers,

Neon lights rotate into rubble on round windows.

Warm air pushes through jagged skylights,

As young lives,

Made of carbon fiber,

Weave the longitude and latitude of cloud shadows.

The scent of books spreads through the concrete columns,

While three or five models step onto waterfalls of light,

Turning the winter color temperature into champagne Mode.

Slanting light slices through the atrium's revolving Staircase,

As the stainless steel handrail drips with the background Music from last year's unopened runway show.

Rust oozes from the peeling graffiti walls,

And lipstick prints in the corners complete the color Reaction.

When clothesline-converted light strips ignite,

A suspended weaving shuttle suddenly conducts Electricity,

All shadows fade away.

Only the newly planted fiddle-leaf figs record in their Memo,

This moment—

Warm sunlight stitches spring's new garments for the Old factory brand.

江城之虹：工业遗产影像志

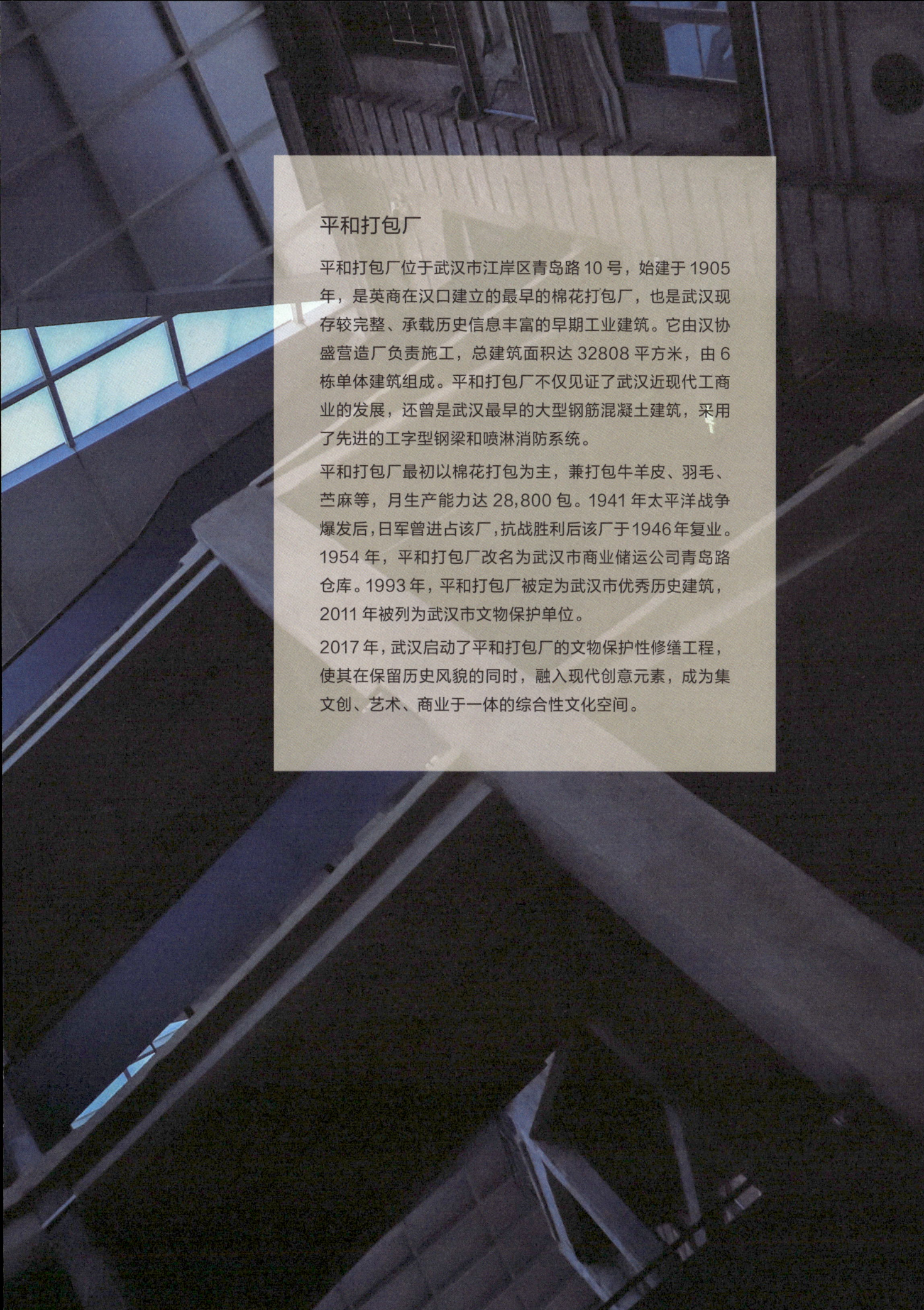

平和打包厂

平和打包厂位于武汉市江岸区青岛路 10 号，始建于 1905 年，是英商在汉口建立的最早的棉花打包厂，也是武汉现存较完整、承载历史信息丰富的早期工业建筑。它由汉协盛营造厂负责施工，总建筑面积达 32808 平方米，由 6 栋单体建筑组成。平和打包厂不仅见证了武汉近现代工商业的发展，还曾是武汉最早的大型钢筋混凝土建筑，采用了先进的工字型钢梁和喷淋消防系统。

平和打包厂最初以棉花打包为主，兼打包牛羊皮、羽毛、苎麻等，月生产能力达 28,800 包。1941 年太平洋战争爆发后，日军曾进占该厂，抗战胜利后该厂于 1946 年复业。1954 年，平和打包厂改名为武汉市商业储运公司青岛路仓库。1993 年，平和打包厂被定为武汉市优秀历史建筑，2011 年被列为武汉市文物保护单位。

2017 年，武汉启动了平和打包厂的文物保护性修缮工程，使其在保留历史风貌的同时，融入现代创意元素，成为集文创、艺术、商业于一体的综合性文化空间。

Pinghe Baling Factory

Located at No. 10 Qingdao Road in Jiang'an District, Wuhan City, Pinghe Baling Factory was established in 1905 and is recognized as the earliest cotton baling factory built by British merchants in Hankou. It stands as a valuable remnant of early industrial architecture in Wuhan, showcasing the well-preserved and richly informative historical structures from that era. Constructed by the HanXiesheng Construction Factory, the factory spans a total floor area of 32808 square meters and consists of six single buildings. Pinghe Baling Factory not only witnessed the evolution of Wuhan's modern industry and commerce, but also marked a significant milestone as the first large-scale reinforced concrete building in the city, featuring advanced I-beam structures and sprinkler fire-fighting systems.

Originally, the factory primarily focused on cotton baling but also processed cattle and sheep skins, feathers, and ramie, achieving a monthly production capacity of 28,800 bales.During the Pacific War in 1941, the Japanese army occupied the factory. It resumed operations in 1946 after the victory in the War of Resistance. In 1954, it was renamed the Qingdao Road Warehouse of Wuhan Commercial Storage and Transportation Company. The factory was designated an outstanding historical building in Wuhan in 1993 and listed as a Wuhan Municipal Cultural Relics Protection Unit in 2011.

In 2017, Wuhan launched a cultural relics protection restoration project for Pinghe Baling Factory, transforming it into a comprehensive cultural space that integrates cultural creativity, art, and commerce while preserving its historical appearance and incorporating modern creative elements.

鹅社
Goose Corporation

鹅社书店艺术馆
Goose Corporation Bookstore & Art Gallery

良有集合·学社艺术·咖啡轻食

鹅社文化传播有限公司
学社艺术 | 建筑概念 | 品牌整合

鹅社书店艺术馆

正在营业
进门请进

营业时间
10:30
22:00

咖啡

学社·艺术·系列·咖啡·酒

鹅社

NEW RED WINE PRODUCE

微醺上新 浪漫必备

红酒&起泡酒务洋酒
限约赠送经典小食

Goose
鹅酒吧

棉絮与代码的复调

红砖缝里流出泛黄的棉花纤维

工字钢梁将灰蒙的世纪折成五线谱

消防喷头顶部的铸铁莲蓬头

正将往事凝结成琥珀色吊灯

阳光切开混凝土肋拱

每一粒骨料都在诉说着历史

那些曾捆扎棉包的钢带

此刻正为全息投影束紧腰身

水磨石地面浮出蒸汽机车压痕

防滑纹路里游弋着数据蝌蚪

昔日的钢架悬停在半空

把空间晾晒成透明的时间胶囊

仓库改装的书店深处

棉花静电正与光纤私语

当年测算体积的滑轨标尺

此刻正丈量着元宇宙的曲率

黄昏漫过琴键似的壁柱

窗棂线脚抖落租界时代沉积的露水

而新安装的玻璃幕墙内侧

3D 打印机正吐出汉江的潮汐

当午夜钟声敲响第七下

消防栓改装的咖啡机开始运转

交错的钢梁成为共鸣箱

打包厂的百年震颤

正被悄悄译作加密的艺术语言

A Polyphony of Cotton Fibers and Codes

Red brick seams bleed time-yellowed cotton threads,

While steel I-beams crease the ashen century into staff lines.

The cast-iron rosettes atop the fire sprinklers condense the past into Amber chandeliers.

Sunlight slices through the ribbed arch of concrete;

Every grain of aggregate tells the story of history.

The steel bands that once bound the cotton bales now cinch holographic Projections in their embrace.

The terrazzo floors bear steam indentations,

As tadpoles of data swim in non-slip lines.

The steel frames of the past hover in mid-air,

Drying the space into a transparent time capsule.

Deep within the warehouse, now a bookstore,

Cotton static whispers with fiber optics.

The slide rule that once measured volume now gauges the curvature of The metaverse.

Dusk spreads over the piano-key-like pilasters,

As the threads of the window panes shake off the dew deposited during Concession era.

Inside the newly installed glass curtain wall,

The 3D printer spews forth the tides of the Han River.

When the clock strikes the seventh stroke of midnight,

The coffee machine modified from the fire hydrant begins to operate.

All the steel beams resonate like boxes,

Silently translating the centuries-old tremor of the packing house into Encrypted artistic language.

江城之虹：工业遗产影像志

宗关水厂

宗关水厂位于武汉市硚口区，始建于 1906 年，1909 年 9 月 4 日正式开始供水。它是武汉最早的自来水厂，武汉人第一次喝到的自来水便来自这里。宗关水厂的前身是商办汉镇既济水电股份有限公司，由浙江商人宋炜臣集资组建，得到了湖广总督张之洞的批准和支持。如今，宗关水厂已发展成为日供水能力为 105 万吨的现代化大型自来水厂，服务人口 200 多万，出厂水质达到欧盟标准。厂区内保留有红砖红瓦的两层小洋楼和酷似欧洲古堡式建筑的送水泵房等百年建筑，其中送水泵房化身为"武汉市自来水事业百年历史展示馆"。

Zongguan Water Plant

Zongguan Water Plant, located in the Qiaokou District of Wuhan, commenced construction in 1906 and officially began supplying water on September 4, 1909. As Wuhan's earliest water treatment plant, it provided the city's first tap water to residents. The plant originated from the Shangban Han Town Jiji Water and Electricity Co., Ltd., founded by Zhejiang merchant Song Weichen, with approval and support from Zhang Zhidong, the governor-general of Huguang. Over the years, Zongguan Water Plant has grown into a modern, large-scale facility with a daily water supply capacity of 1.05 million tons. It now serves over 2 million people, maintaining water quality that meets European Union standards.The site preserves century-old structures, including: a two-story red-brick building; a castle-like delivery pump house, now repurposed as the "Centennial History Exhibition Hall of Wuhan Waterworks."

水纹年轮

红砖在离心机里沉淀成钟表零件
穿越百年的水压仍在青铜阀门上
雕刻宋体字的年轮
当泵房拱顶垂落为量杯
湖广总督的批文正在反渗透膜上
析出碳酸钙的诗行
送水泵房将潮声砌成古堡
条石计算水分子量程

铸铁管道裂开第三万道缝时
离心机正在翻译欧盟的水语
而消毒池底的青苔始终记得
二十四万张陶碗碎裂的弧度
百年展示馆的玻璃幕墙内
水塔阴影正与氯离子起舞
混凝池里沉睡的矾花
在二十一世纪的浊度仪上
开出宣统元年的结晶

红瓦小楼把水压换算成心跳
值班表在石英砂滤层中持续发酵
当出厂水穿过一九〇九年的量杯
所有数字都在铸铁管道里裂变生长

夜色漫过沉淀池边缘时
泵房突然吐出整条汉江
那些被软化处理的历史
正以微米级的精度
重新编织一座城市的毛细血管网

The Rings of Water Ripples

Red bricks precipitated into clock parts in the centrifuge.

The water pressure that has lasted for a hundred years still carves the rings in Song characters on the bronze valves.

When the vault of the pump house drops down as a measuring cup,

The governor-general of Huguang's approval document precipitates calcium Carbonate verses on reverse osmosis membranes.

The pump house builds the sound of the tide into an ancient fortress,

Every stone is calculating the molecular weight of water.

When the cast iron pipe cracks its 30,000th seam,

Centrifuges are translating the water language of the European Union,

And the moss at the bottom of the sterilized pool always remembers the Curvature of 240,000 broken bowls.

Inside the glass walls of the Centennial Exhibition Hall,

The shadows of the water towers tango with the chlorine ions.

The sleeping alum flowers in the coagulation tank ,

On the turbidimeter of the 21st century,

Crystallized in the first year of the Xuantong era.

The red-tiled building converts water pressure into heartbeats,

The duty roster continues to ferment in the quartz sand filter.

When the factory water passes through the measuring cup of 1909,

All the numbers are fissioning and growing in the cast iron pipes.

As night spreads over the edge of the settling tanks,

The pump house suddenly vomits out the entire Han River.

The histories that have been softened and treated are now,

With micron-level precision,

Reknitting the capillary network of a city.

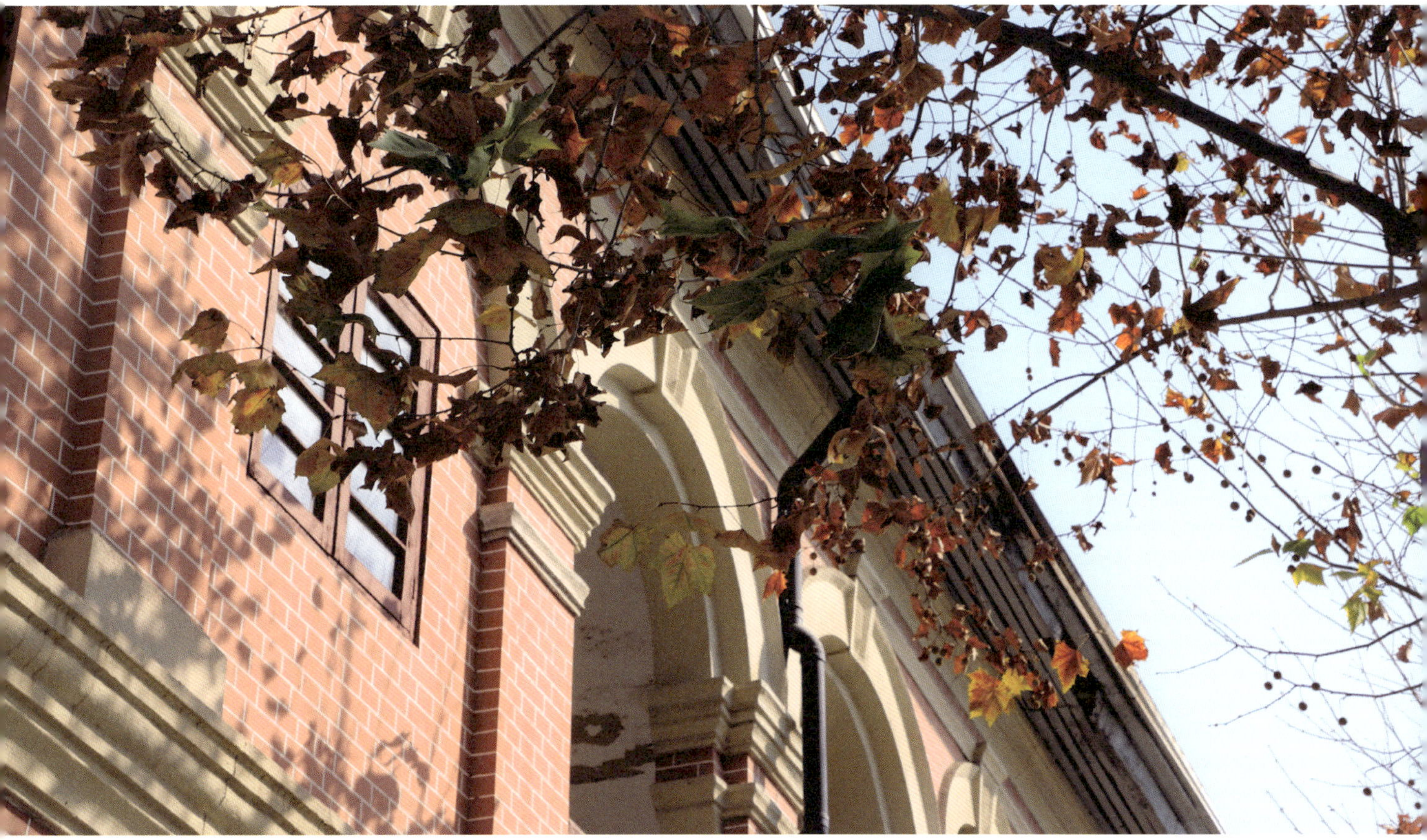

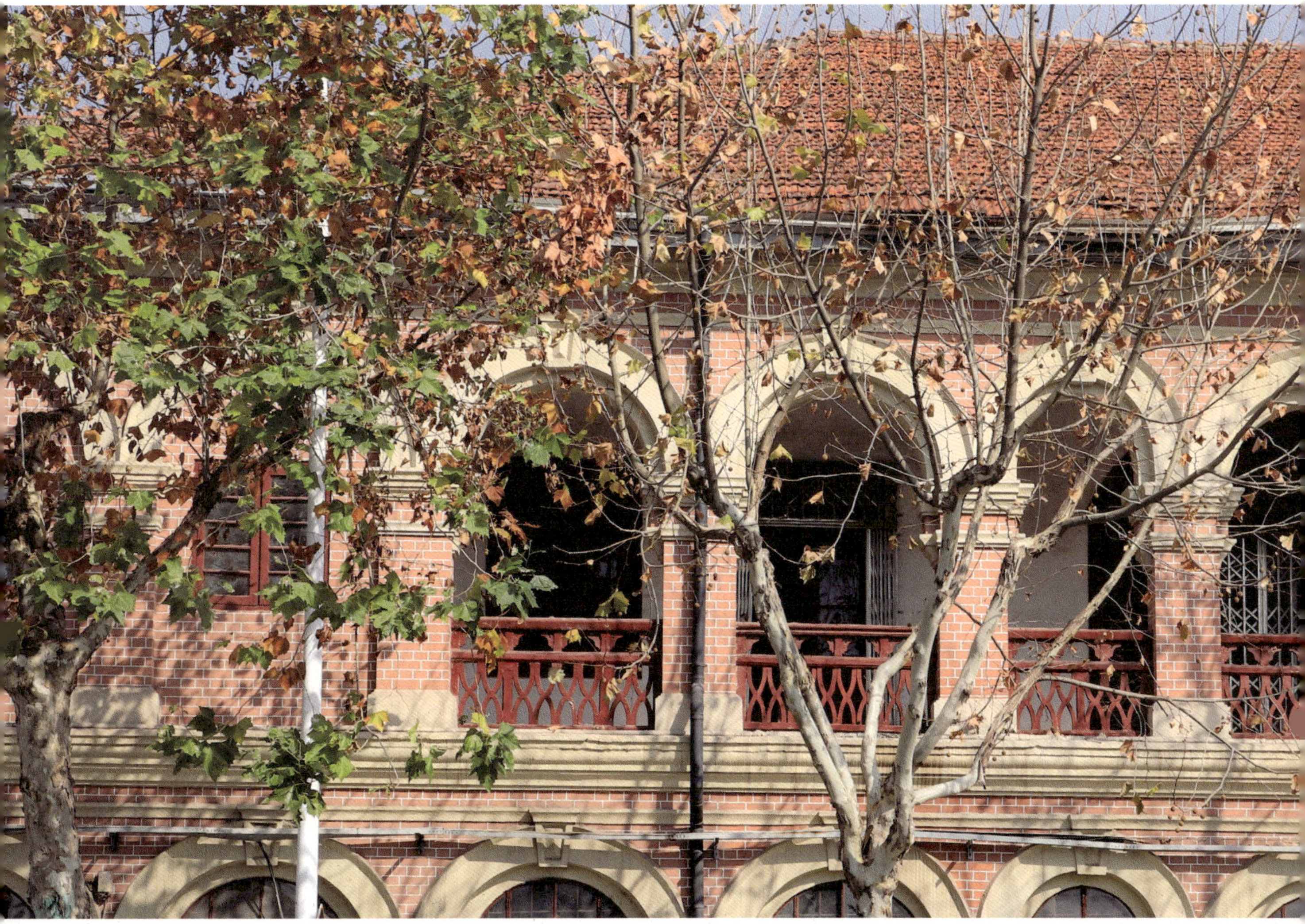

武汉良友红坊文化艺术社区

武汉良友红坊文化艺术社区位于江岸区汉黄路 32 号，前身为 20 世纪 60 年代的湖北省皮毛厂，20 世纪 90 年代被用作建材市场。2017 年，结合厂区原有建筑，政府与设计团队以公共艺术手法和城市再生理念进行改造，最大限度保留历史记忆和工业遗产。园区占地面积约 8 万平方米，总建筑面积约 6 万平方米。武汉良友红坊文化艺术社区保留了整体结构，强调场地记忆，如红砖墙、坡屋顶和红瓦屋面等。设计团队巧妙运用"保留与重塑"手法，在尊重原有风貌的基础上加入创新元素，使空间既富有历史韵味又不失现代感。

Wuhan Liangyou Hongfang Culture and Art Community

Wuhan Liangyou Hongfang Culture and Art Community is located at No. 32 Hanhuang Road in Jiang'an District. Once the Hubei Fur Factory in the 1960s, the site later served as a building materials market in the 1990s. In 2017, the government and a design team transformed the original factory buildings using public art techniques and urban regeneration concepts to maximize the retention of historical memory and industrial heritage. Covering an area of approximately 80,000 square meters, with a total building area of around 60,000 square meters, Wuhan Liangyou Hongfang Culture and Art Community retains its overall structure, emphasizing site memory with features like red brick walls, sloping roofs, and red-tiled roofs. The design team skillfully employed a "preservation and remodeling" strategy that respects the original style while integrating innovative elements, giving the space a historical flavor intertwined with modernity.

时光的褶皱

红砖在皮肤下凝固成血

六十年代的铁钉依然悬垂于

混凝土的静脉中

那些被切割的棱角

在阳光里重新发芽

长出玻璃般透明的菌丝

推土机的手臂已沉入地基深处

钢铁撑开穹顶且

正沿着钢筋的纹路编织新茧

有人把月亮敲打成

一组不锈钢的密码

锈蚀的齿轮

突然在视网膜上转动起来

斜坡屋顶收留着所有迁徙的光

文字在砖缝间蜕皮

落地成铅字雨

羽毛球划过天际

快门声惊飞灰鸽一群

瓦片从未忘却

自己曾是某座山脉年轻的锁骨

书架正在吞噬寂静

混凝土长出气根

缠绕所有未完成的命名

当焊枪点亮第六十次日落

我们终于学会

在铆钉与诗句的缝隙里

豢养一整个世纪的潮声

A Wrinkle in Time

Red bricks coagulate into blood beneath the skin,

With iron nails from the 1960s still dangling in the veins of
Concrete.

Those cut angles germinate anew in the sun,

Growing glass-like transparent hyphae.

The bulldozer's arm has sunk deep into the foundations,

While steel opens the dome,

Weaving new cocoons along the grain of steel.

Someone hammered the moon into a set of stainless steel
Passwords,

Rusted gears suddenly begin to rotate upon the retina.

The sloping roof captures all the migrating light,

As words molt between brick cracks,

Landing in a rain of lead.

The shuttlecock arced across the horizon,

The sound of the shutter startled a flock of gray pigeons.

The tiles always remember that they were the young
Collarbones of some mountain ranges.

The bookshelf is devouring the silence,

As concrete grows aerial roots that entwine with all the
Unfinished naming.

When the torch lights the sixtieth sunset,

We finally learn in the crevices of rivets and verse to keep a
Whole century's worth of tides.

站前花街

站前花街的正式名字叫车站前路，位于汉阳火车站旁。这条不足 300 米长的小路，原本是汉阳枕木防腐厂的专用货运铁路，一头连着汉阳大道，一头伸向京广线上的汉阳火车站，汉阳枕木防腐厂搬走后，铁路被废弃，路面变得宽敞，这才成了居民的日常步道。

走进站前花街，映入眼帘的是一面旧时的砖墙，装饰有各个时期的列车指示牌，加上改造后铺满石子的钢轨小路，在这里拍出的照片自带复古滤镜。漫步在钢轨小路两侧，感觉像乘坐时光机回到了过去。附近的居民每天都会在树荫下择菜聊天、吃饭喝茶，成了站前花街一道独特的风景线。顺着钢轨小路往前走，会发现路边和墙上长满了各色的花卉植物，有绣球花、栀子花、石榴花等，甚至在很多角落还能遇见一幅幅漂亮的画作。原来，附近的居民都喜欢种花和画画，他们把家里的花花草草搬到了家门口，原本破旧的墙面也渐渐被街坊们当成白纸用来作画。钢轨两旁也因此摇身一变，成了漫画般独一无二的景色。

2024 年 5 月，站前花街新增了两节绿皮火车厢，设置在距站前花街入口 50 米处。这两节绿皮火车厢成为最为醒目的标识景观。根据规划，绿皮火车厢将被打造为咖啡屋或阅读角等多元化的公共服务空间，方便游客休息、打卡。

Stationfront Flower Street

Stationfront Flower Street, officially named Chezhanqian Road, lies beside Hanyang Railway Station. This under-300-meter-long lane was originally a dedicated freight railway for the Hanyang Railroad Tie Treatment Plant, connecting Hanyang Avenue at one end and extending toward Hanyang Railway Station on the Beijing-Guangzhou Line at the other. After the plant's relocation, the railway was abandoned, and the widened path gradually became a community walkway for residents.

Upon entering the street, an aged brick wall adorned with historical train signage greets visitors. Combined with the renovated gravel-paved rail tracks, photos taken here naturally exude a vintage charm. Strolling along the tracks evokes the sensation of traveling back in time. Locals gather daily beneath shaded trees to prepare vegetables, chat, and enjoy meals—a distinctive neighborhood tableau. Walking further along the tracks, vibrant blooms and murals emerge: hydrangeas, gardenias, pomegranate blossoms, and other flowers blanket walls and pathways, while artistic paintings grace unexpected corners. Residents, passionate about gardening and art, have transformed their doorsteps into floral displays and repurposed weathered walls as canvases. Thus, the railside area has blossomed into a comic book-like wonderland.

In May 2024, two green-painted train carriages were added 50 meters from the street's entrance, now serving as its most striking landmarks. According to urban plans, these carriages will be converted into versatile public spaces—such as cafes or reading lounges—providing rest areas for visitors to relax and capture memorable photos.

枕木在花影中返青

锈蚀的钢轨静脉曲张般
蜿蜒成约三百米长的时间胶囊
枕木里渗出的防腐剂
正被绣球花的根系搅拌

信号灯锈成怀表零件
悬在静脉注射架般的砖墙上
石子路基上遗落的货单编号
被栀子花香重新装订成日历

择菜声校准蒸汽钟
茶渍在铁轨接缝处拓印年轮
绿皮火车厢卡进时空褶皱
铆钉在咖啡渍里发芽新生
书页掀起的煤烟味中
墙画从砖缝里牵出藤蔓

石榴花炸开的刹那
所有防撞杆都退化成画框
而爬山虎正沿着铁轨刻度
把锈色换算成叶绿素

暮色浸透车厢接缝
整条花街逆向而行
花盆里的波斯菊举起信号旗
风声扬起
一代人的青涩梦想
与二十一世纪的霓虹光谱
在枕木里交接示意
原来是向日葵跑进了秋天

The Sleepers are Greening in the Shadow of the Flowers

The rusted rails are like varicose veins,

Winding into a time capsule about three hundred meters long.

The preservative oozing from the sleepers is stirred by the Hydrangea's root system.

Signal lights,

Rusting into pocket watch parts,

Suspended on brick walls like IV racks.

The numbers of the bills of lading rambling on the gravel roadbed
Are rebound into calendars by the scent of gardenias.

Residents calibrate steam clocks with the sound of preparing Vegetables.

Tea stains mark the seams of the railroad tracks.

Green-painted train carriages are stuck in the spacetime Folds.

Rivets sprout anew in coffee stains.

The pages of a book lift the smell of soot,

Wall paintings pull out vines from the cracks in the bricks.

The moment the pomegranate blossom bursts,

All bumping posts are degenerating into picture frames,

And the ivy is following the railroad track scale,

Converting rust colors into chlorophyll.

As twilight soaks the seams of the carriages,

The whole Stationfront Flower Street travels in reverse.

Persion daisies in pots raise signal flags,

The sound of the wind lifts up a generation's youthful dreams,

To exchange nods with the neon spectrum of the twenty-first Century in sleepers.

Turns out it was the sunflowers that ran into the fall.

大智无界·空中小镇创意产业园

大智无界·空中小镇创意产业园位于武汉市江岸区，前身是 1961 年成立的武汉市无线电厂。武汉市无线电厂曾以"长江音响"这一拳头产品而闻名。2017 年，江岸区获批国家级双创示范基地后，对厂内 7 栋老旧厂房进行改造升级。园区总建筑面积约 5 万平方米，由多栋建筑围合而成。园区主要打造创意设计、城市展厅、众创空间、时尚生活、文化创意、科技创新等新兴业态。

Dazhi Boundless-Air Town Creative Industrial Park

Dazhi Boundless-Air Town Creative Industrial Park is situated in Jiang'an District, Wuhan City, originated from the Wuhan Radio Factory established in 1961.This factory once gained fame for its flagship product, the "Yangtze River Audio." Following Jiang'an District's designation as a National Mass Entrepreneurship and Innovation Demonstration Base in 2017, seven historic workshop buildings within the complex underwent comprehensive renovation and modernization.The park now encompasses approximately 50,000 square meters, consisting of several buildings. It emphasizes creative design, urban exhibition halls, crowdsourcing spaces, fashionable life, cultural creativity, technological innovation, and other emerging industries.

信号与场域

晶体管在混凝土裂缝中发芽
悬空的电缆正触摸月光
孤独的玻璃栈道，任霓虹
在氧化铝窗框上调试纯度
防辐射铅门虚掩成书店
示波器外壳破茧成环形吧台
车间里摆出了后现代的影像
铆钉的螺纹仍在测绘心跳

阴影被钢格栅筛成细沙
她的指尖掠过通风管道——
七十年代的热浪突然苏醒
在镀锌波纹管表面凝结水珠
当老车床切削的寂静
迎面撞上展厅的照片
慌乱中
他抓住一截电流

总控台锈蚀的按钮
努力将消失的电磁波
折叠成咖啡拉花的涟漪
而此时此刻的数据
正流过防火堤
在服务器集群的褶皱里
织出第一道双向导通的光路

The Signal and the Field

Transistors germinate in concrete cracks,

Suspended cables touch moonlight,

Lonely glass walkway permits neon to debug purity
On aluminum oxide window frames.

Radiation-proof lead doors vaporize into bookstores,

Oscilloscope casings grow into a circular bar,

While the workshop adopts a postmodern visage,

Threads of rivets mapping heartbeat frequencies.

Shadows sift into sand through steel grills.

Her fingertips skim along the ventilation ducts—

The heat wave of the seventies suddenly awakens,

Condensation beads form on the surface of galvanized
Bellows.

When the silence cut by old lathes collides head-on
With exhibition hall photographs,

In panic,

He seizes a current's fragment.

The rusted buttons of the console attempt to fold the
Vanishing electromagnetic waves,

Transforming them into ripples of coffee pulls.

And right now,

Data pours through the firewall,

In the folds of the server cluster,

Weaving the first bi-directional paths of light.

华中小龟山金融文化公园

华中小龟山金融文化公园位于武汉市武昌区，面积约88.09亩。这里曾是湖北省电力建设第一工程有限公司于20世纪70年代修建的后方生产基地，如今已变身为华中首个以金融为主题的产业园区。园区内坐落着22栋单体建筑，均由老旧厂房"修旧如旧"而成，红砖青瓦的外观怀旧气息满满。

华中小龟山金融文化公园不仅是一个金融产业园区，更像一座闹中取静的"金融小镇"。它结合了自然与文化元素，打造了多个景观节点，如龙门秀场，即为多元化、可运营的艺术活动场地，可举办交流、聚会等各种活动，是整个园区的活力核心。此外，小龟山作为项目的生态之心，通过设计还原山体坡地，打造不同高差的公园绿地，让人们得以再次拥抱小龟山，实现自然与人文的和谐共生。

Central China Xiaoguishan Financial and Cultural Park

Central China Xiaoguishan Financial and Cultural Park is located in Wuchang District, Wuhan City, covering an area of about 88.09 mu. This park was once the rear production base established by the Hubei No.1 Electric Power Construction Co.,Ltd., in the 1970s. It has now been transformed into the first financial-themed industrial park in Central China. The park features 22 single buildings, all of which are historical factory structures revamped to retain their nostalgic charm, characterized by red bricks and green tiles.

Central China Xiaoguishan Financial and Cultural Park is not just a financial industrial park; it also serves as a "financial town" nestled within a tranquil neighborhood. The park seamlessly integrates natural and cultural elements, creating various landscape nodes such as the Longmen Showground. This venue has evolved into a versatile space for art activities, hosting exchanges, gatherings, making it the vibrant heart of the entire complex. Additionally, the Xiaoguishan acts as the ecological core of the project, designed to restore the mountain slopes and develop parks with varied elevations, allowing visitors to reconnect with Xiaoguishan and fostering a harmonious relationship between nature and humanity.

金融季风

金融季风穿过红砖褶皱

钢索与龙门吊作别

桁架骨骼覆满玻璃的羽鳞

红砖墙体内

七十年代的电缆

正把数据流编织成竖琴

龙门秀场的钢梁悬垂为琴弦

招商手册里未拆封的铆钉

在咖啡杯沿幻化成数据

金融算法攀爬藤蔓的韵脚

将现金流换算成叶绿素

当混凝土立柱裂开第三十一道缝

财务报表拨转年轮

会计师用标尺丈量钢窗的斜阳

在光影里藏进复利公式

旧变电箱改装的雕塑

利落吞吐着跨国汇率的季风

当红移光谱漫过资产负债表

山坳便以复绿率的名义

将每寸钢渣换算成碳汇额度

而老塔吊的阴影始终在计算

资本与山岚的兑换汇率

The Financial Monsoon

The financial monsoon sweeps through the folds of red
Bricks,

As gantry cranes unload steel cables,

The truss skeletons growing glassy scales.

Within the red brick walls,

Cables from the 1970s weave a data flow that forms a
Delicate harp,

With the steel girders of the Longmen Showground
Dangling like strings.

Unopened rivets in investment brochures,

Dissolve into data on the rims of coffee cups.

Financial algorithms ascend the rhyming tendrils of
Creepers,

Transforming cash flow into chlorophyll.

Concrete columns fracture at their thirty-first seam,

As financial statements begin to secrete annual rings.

Accountants measure the slanting sunlight streaming
Through steel windows with rulers,

Each frame of light conceals a compound interest formula.

The sculpture repurposed from an old transformer box
Inhales the monsoon of cross-border exchange rates.

As redshifted spectra flood balance sheets,

The valley,

In reforestation rate's name translates each slag
Fragment to carbon credits,

While the old tower crane's shadow forever calculates
The exchange rate between capital and mountain mists.

四美塘铁路遗址文化公园

四美塘铁路遗址文化公园位于武汉市武昌区四美塘，前身为武九铁路北环线武昌北站及机修车间段。这里曾是南北货物运输的动脉，见证了中国铁路系统从萌芽到繁荣的变迁。公园保留了铁轨、枕石、信号灯、火车头、绿皮火车厢等铁路元素，创造了有特色的景观。园内还复原了老火车站、徐家棚历史建筑、铁路驿站等。四美塘铁路遗址文化公园是长江百里生态文化长廊的重要节点，于2023年11月向公众开放。

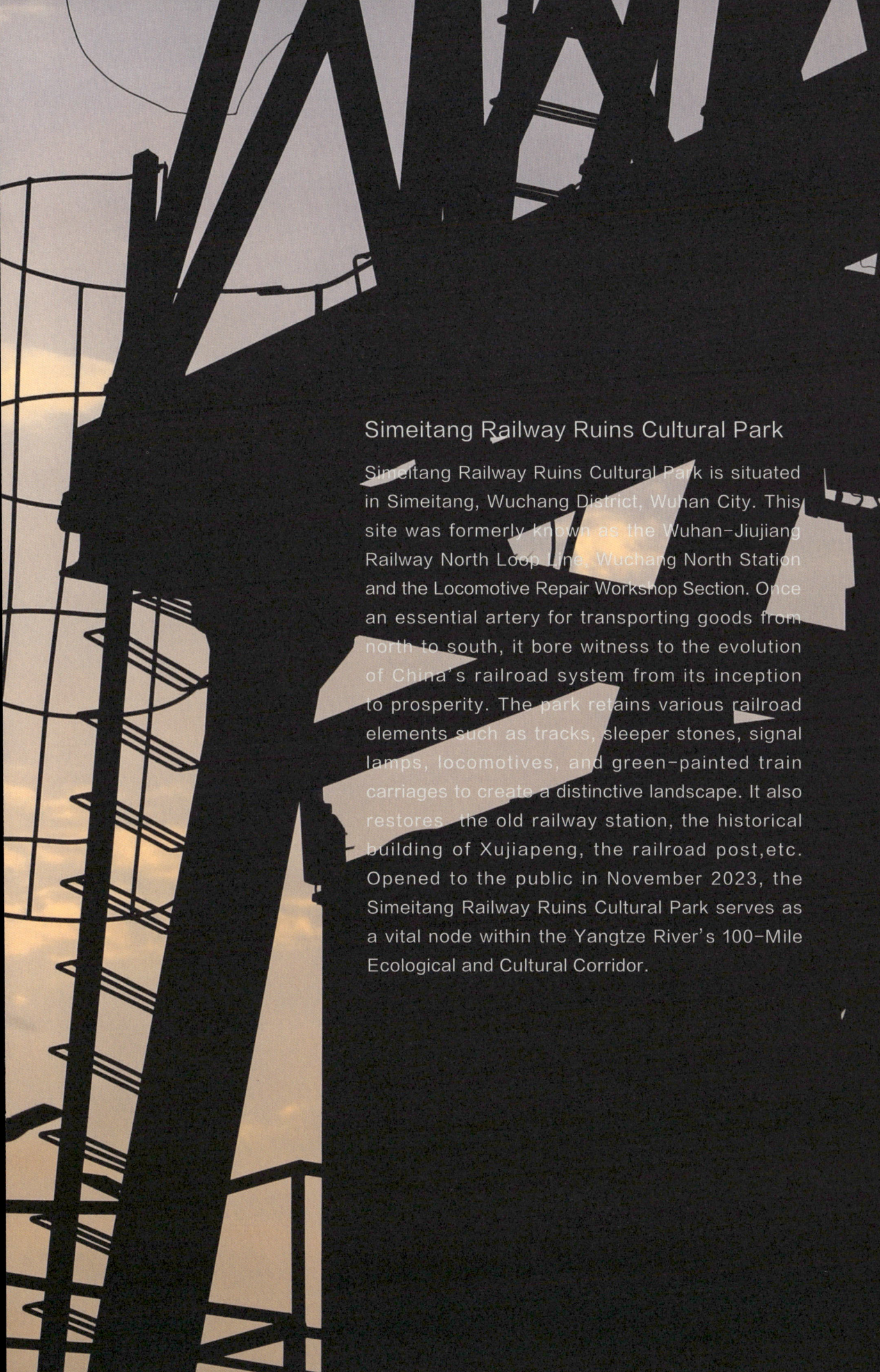

Simeitang Railway Ruins Cultural Park

Simeitang Railway Ruins Cultural Park is situated in Simeitang, Wuchang District, Wuhan City. This site was formerly known as the Wuhan–Jiujiang Railway North Loop Line, Wuchang North Station and the Locomotive Repair Workshop Section. Once an essential artery for transporting goods from north to south, it bore witness to the evolution of China's railroad system from its inception to prosperity. The park retains various railroad elements such as tracks, sleeper stones, signal lamps, locomotives, and green-painted train carriages to create a distinctive landscape. It also restores the old railway station, the historical building of Xujiapeng, the railroad post, etc. Opened to the public in November 2023, the Simeitang Railway Ruins Cultural Park serves as a vital node within the Yangtze River's 100-Mile Ecological and Cultural Corridor.

枕石上的刻度

锈蚀的铁轨在二月的草甸上延伸成

时光的游标卡尺

蜿蜒生出青苔的枕石

还残留着二十世纪蒸汽的余温

信号灯在悬铃木枝丫间眨眼

将黄昏调制成扳道工的萤火

当年的洼地

此刻正托着桁架的倒影

泛起的涟漪

将货运时刻表揉成月影

混凝土桥墩裂开缝隙

野蔷薇窜改调度密语

废弃的转辙器突然偏转角度

跳舞的人群沿着道岔

滑入月光校准的慢板声里

那些被截断的钢轨

正在发芽——

虹桥从焊缝里拱起

货运站台褪色的编号

正被轮滑鞋画出新的坐标系

当末班有轨电车的回音

卡进内燃机车车轮的齿隙

沉睡的转盘道突然通电

所有铁质记忆开始环行

锈红血管沿着江岸蜿蜒

不分昼夜

运送整座城市的星群

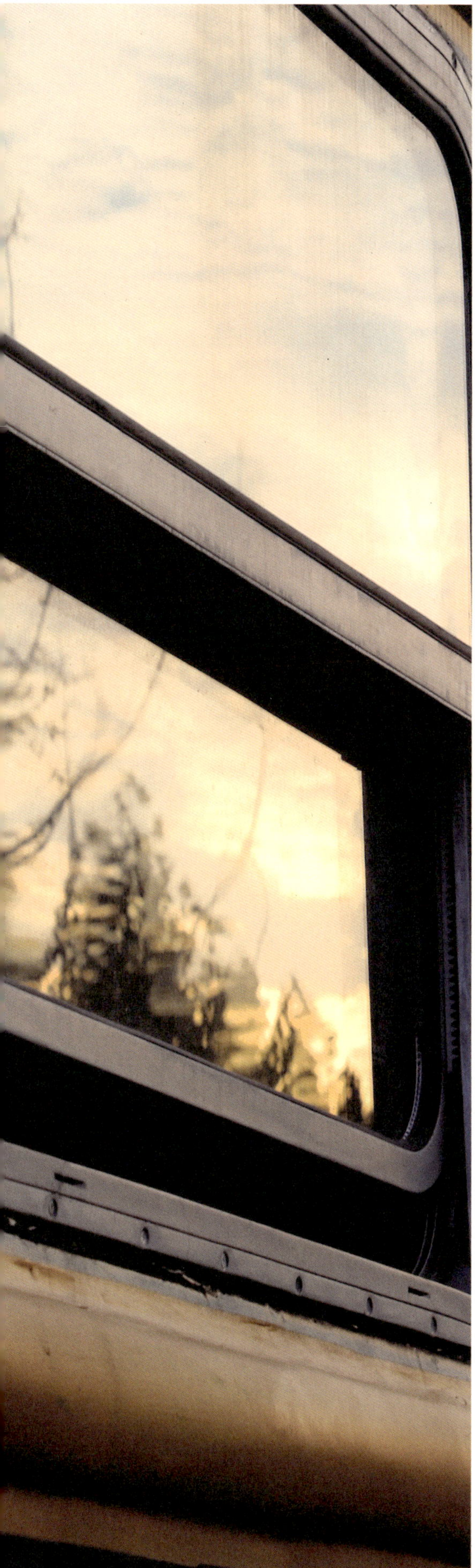
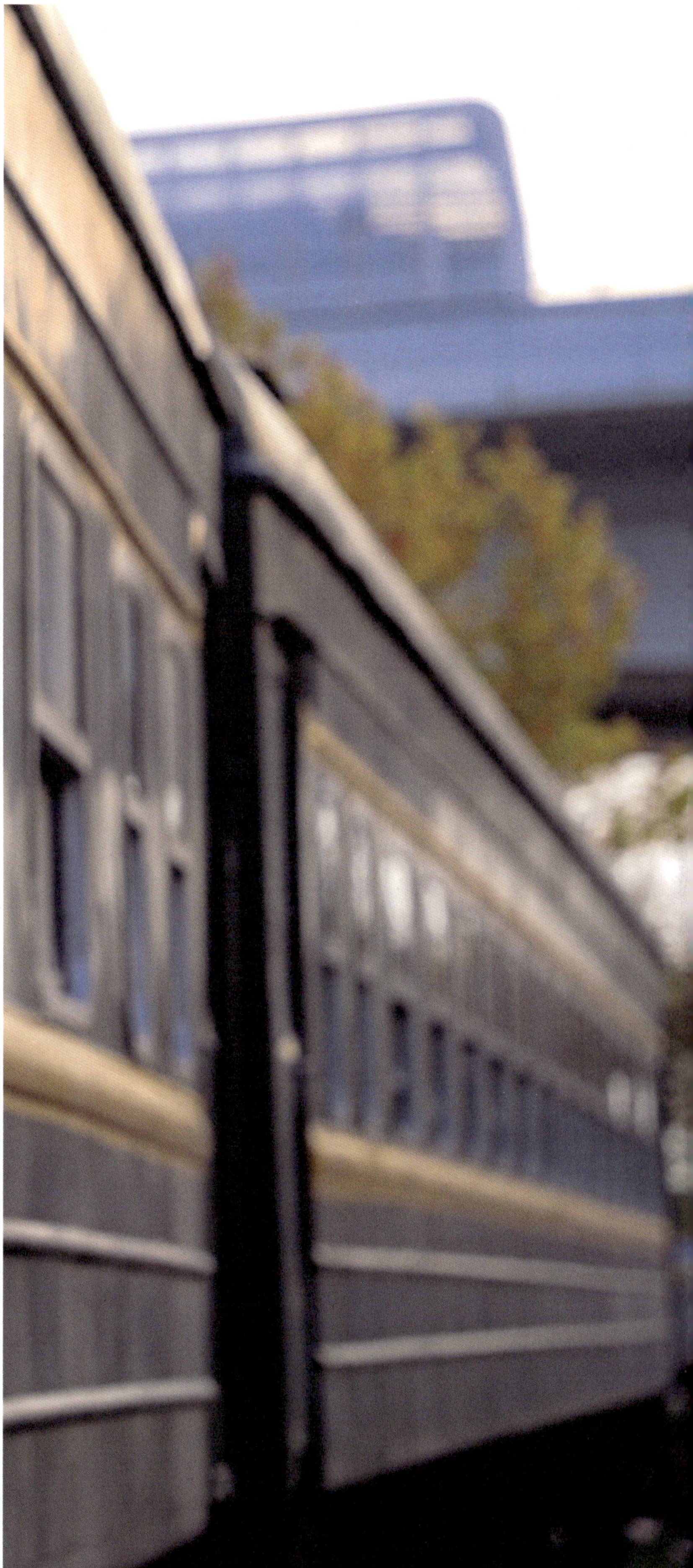

The Scale on the Sleeper Stone

Rusty railroad tracks extend across the February meadow
Becoming a vernier caliper of time.
Sleeper stone with moss still retaining the residual warmth
Of twentieth-century steam.

Signal lamps wink in the branches of plane trees,
Modulating dusk into a switchman's glow.
The depression of those years now holds the reflection of
Trusses,
The ripples arising crumple the freight timetable into a
Moon shadow.

Where concrete piers crack,
Wild roses tamper with the dispatch code.
Abandoned rutters are suddenly deflected,
Letting dancing crowds along the switchbacks slide into a
Moonlight-calibrated slow motion.

Those truncated rail are sprouting—
Rainbow bridges arch from welded seams.
The faded numbers of freight platforms are traced by roller
Skates to a new coordinate system.

As the echoes of the last tram get stuck in the teeth of
Internal combustion locomotives,
The sleepy turnpike suddenly electrifies,
All the iron memories begin to circulate.
Rusty red veins snake along the riverbank,
Day and night,
Carry the stars of the city.

武钢云谷·606 产业园

武钢云谷·606 产业园位于武汉市洪山区，北临团结大道，东临青王路，南临工业园路，西临三环线，紧邻入选"全球最美建筑"的武汉火车站。园区原为始建于 1954 年的武汉冶金设备制造厂，占地 606 亩，是武汉主城区内少有的工业遗存相对集中、风貌相对完整的区域。这里曾是武汉市工业文明的象征，如今已摇身一变成为文创和科创园区，成为武汉市重点建设的文创项目。园区内分布有多处工业遗迹，如水塔、龙门吊等，这些元素赋予了园区鲜明的工业特色。2022 年 3 月，武钢集团启动了首发区项目建设，按照"修旧如旧、空间织补、功能完善"的理念，改造了 13 座特色建筑。

Wugang Yungu · 606 Industrial Park

Wugang Yungu · 606 Industrial Park is located in Hongshan District, Wuhan City, bordered by Tuanjie Avenue to the north, Qingwang Road to the east, Industrial Park Road to the south, and the Third Ring Road to the west. It is adjacent to Wuhan Railway Station, which has been selected as one of the "World's Most Beautiful Buildings." The park occupies the site of the former Wuhan Metallurgical Equipment Manufacturing Plant, originally established in 1954. Covering 606 mu, it represents one of the few areas in Wuhan's main urban district with relatively concentrated and well-preserved industrial heritage. Once a symbol of Wuhan's industrial civilization, it has now transformed into a cultural and technological innovation hub, serving as a key municipal cultural development project. Numerous industrial relics are distributed throughout the park, including water towers and gantry cranes, which imbue the area with distinct industrial characteristics. In March 2022, Wugang Group initiated the construction of the pilot zone. Guided by the principles of "restore the old as the old, spatial reconnection, and functional enhancement," the project revitalized 13 distinctive buildings.

冶金术的转译

龙门吊俯身成为青铜雕塑时
一九五四年的钢水正沿着光缆奔涌
在服务器集群里重构结晶
冷却池浇筑的直播间
霓虹将冶金图谱转译成像素雨

水塔把黄昏蒸馏成液态字节
储存在环形走廊的玻璃容器里
曾锻打火种的汽锤群
此刻托举全息投影键盘
将淬火构件的余温编码成弦理论

钢渣山脊生长硅基苔藓
氧化铁在防爆墙上投射赭红滤镜
老式车床切削的寂静
突然通电
数据流沿着铸铁法兰盘
漫过三十二座厂房的神经突触

穿堂风掀起防尘帘

露出桁架拼接的星轨

质检钢印在混凝土墙上洇成涂鸦底纹

而断齿的齿轮咬住某段电磁波

将午夜的金属疲劳数值

换算为晨雾的刷新频率

当第一缕光切开行车梁

锈蚀的滑轨导通电流

所有阴影坍缩成注释

唯有遗留在钢模上的指纹

正与新打印的二维码

完成一场跨世纪的晶相重组

Transmutation of Metallurgy

When the gantry crane stoops into a bronze sculpture,

Molten steel from 1954 surges through fiber optics reconstructing
Crystals in server clusters—

The cooling pool casts a live-streaming chamber as neon
Translates metallurgical diagrams to pixel rain.

The water tower distills dusk into liquid bytes stored in glass
Vessels along annular corridors.

Pneumatic hammers that once forged sparks now lift holographic
Keyboards,

Encoding the residual heat of thermal-quenched components into
String theory.

Silicon-based moss creeps over slag ridges while iron oxide
Projects ochre filters on blast walls.

The silence lathed by old machines suddenly electrifies—

Data streams course through cast-iron flanges,

Flooding synapses across thirty-two factory buildings.

A through-draft lifts the dust-proof curtain,

Revealing star trails welded from trusses.

Inspection stamps bleed into graffiti on concrete walls.

A broken gear clenches stray radio waves,

Converting midnight's metal fatigue into the refresh rate of morning
Fog.

When dawn's first light slices the crane beam,

Rusted tracks conduct currents—

All shadows collapse into footnotes.

Only fingerprints lingering on steel molds now converse with
Freshly printed QR codes,

Their murmurs completing a cross-century realignment of
Crystalline phases.

143

青山区红房子

青山区红房子采用苏联建筑风格，均为 3 层，楼间距为 15 米，室内空间宽大。从空中俯瞰，其布局呈"囍"字形。红房子的门牌号采用"X 街坊 X 门 X 户"的形式统一编排；街坊内设有篮球场、中心凉亭等公共设施。这里不仅是武钢职工的居住区，更是他们生活和情感的寄托。如今，部分红房子经过修缮改造，成为创意设计中心等文化地标。

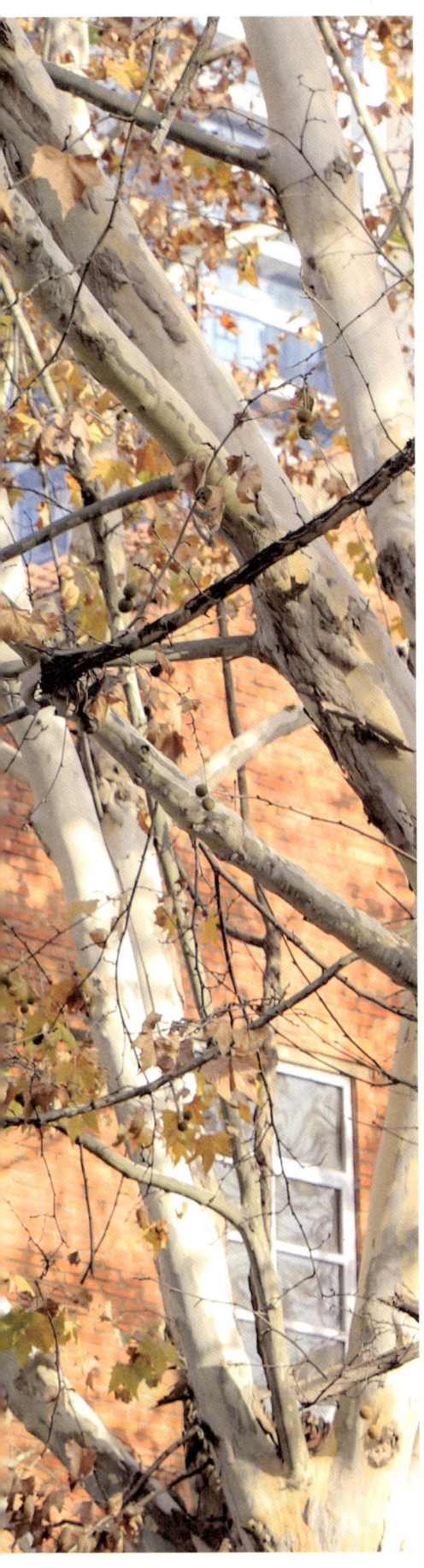

The Red Houses in Qingshan District

The Red Houses in Qingshan District feature Soviet-style architecture, all three stories high with 15-meter building spacings and spacious interior rooms. Aerial views reveal their layout forming a Chinese character "囍" (double happiness). Doorplates follow a unified numbering system: "X Street Block, X Gate, X Household." Public facilities like basketball courts and central pavilions are integrated within the blocks. These complexes served not only as residential quarters for Wuhan Iron and Steel workers, but also as emotional anchors for their lives. Today, renovated sections have been transformed into cultural landmarks such as creative design centers.

砖红色的心脏

三十二万块红砖在经纬线上编织指纹

每一道灰缝都是时间的毛细血管

街坊门牌在锈蚀的钢钉下生长为胎记

信箱里未寄出的信件静静发酵

铁皮屋顶将上世纪的雨水酿成琥珀

篮球场裂成两片肺叶

仍在风中交换着青春的回声

凉亭里

一盘未下完的棋局长出青苔

车马相（象）仕（士）在砖缝间迷途

而老槐树的年轮还在搬运炼钢炉的潮汐

黄昏把苏式拱窗拓印成蝴蝶标本

楼道里浮动的铁锈

仍在测绘体温的曲线

当脚手架剖开砖墙的胸腔

钢筋正以另一种形态苏醒

被岁月氧化的筋骨在玻璃幕墙背面

继续编织新的经纬

月光浸透的砖缝间

有人用霓虹续写未完的梦想

每扇亮灯的窗口都是密码

解码时分

混凝土的纹路里

翻涌着整座钢铁森林的脉搏

The Brick-red Heart

Three hundred and twenty thousand red bricks weave the
Fingerprints of time into their warp and weft,
Every mortar seam —
A capillary of the time.
Neighborhood door plates age like birthmarks beneath
Rusted steel nails,
While unsent letters fester in the mailboxes,
Tin roofs brew last century's rain into amber.

The basketball court cracks like two lungs,
Still exchanging echoes of youth carried on the breeze.
In the gazebo,
Moss blankets an unfinished game of chess,
While chariots, horses, elephants and guards are lost in
The crevices of the bricks.
The old acacia tree bears witness,
Its growth rings carrying the tides of the steel furnace.

Dusk paints the Soviet-style arched windows,
Transforming them into butterfly specimens,
Floating rust in stairwells still surveys curves of body heat.
When scaffolding cuts open the chest of the brick wall,
Reinforcing steel awakens in a new form.
The oxidized tendons along the glass curtain wall continue
To weave a fresh tapestry of history.

In the moonlight-drenched brick cracks,
Someone continues to sketch unfinished dreams with neon.
Every illuminated window serves as a code;
When deciphered,
It reveals the texture of concrete,
And the pulse of an entire steel forest springs forth.

403 国际艺术中心

403 国际艺术中心位于武汉市武昌区百瑞景中央生活区，由原武汉锅炉厂编号 403 的双层车间工业遗址改造而成，总面积达 3429.31 平方米。403 国际艺术中心的外观保留了工业遗址的复古风格，内部则融合了现代设计元素，如超大钢筋框架的挑梁和大面积玻璃屋顶，营造出一种超高透明空间，给每个步入者以极大的视觉享受。403 国际艺术中心不仅是一个文化艺术的孵化地，更是武汉人关于老武锅的记忆所在。

The 403 International Art Center

The 403 International Art Center is situated in the central living area of Bairuijing, Wuchang District, Wuhan City, this center occupies the former double-story workshop of the Wuhan Boiler Factory, encompassing a total area of 3429.31 square meters. The exterior of the 403 International Art Center retains the vintage charm of its industrial past, while the interior embraces modern design elements, featuring oversized steel frame beams and expansive glass roofs that create an ultra-transparent environment, inviting visitors to enjoy the views. The center serves not only as an incubator for culture and art but also as a space where the memories of the Wuhan Boiler Factory are held by the people of Wuhan.

钢的复调

十二米挑梁切割天穹

铆钉在记忆的断层生长出铜绿枝丫

红砖墙体内

四十年前的蒸汽锤

仍在敲打光的韵脚

齿轮悬停在半空

化作青铜编钟的喉舌

淬火的铁屑再度结晶

青年们用剧本测量钢铁的延展度

抑扬顿挫间

台词在玻璃穹顶折射出

六十四种光谱

旋转楼梯螺旋着上升的烟囱

台阶镶嵌半枚焊点

当月光浇筑钢水

混凝土立柱便裂出紫藤的静脉

铸铁横梁长出的年轮

在镜头快门声中轻轻震颤

午夜剧场启幕时

钢筋骨骼流淌出诗句

被遗弃的吊钩挽起天鹅绒幕布

焊接缝里渗出的不是铁锈

而是不愿氧化的青春

空荡荡的剧场里

铜阀门仍在吞吐潮汐

每个螺栓孔都成了透视窗

窥见昔年淬火的黄昏

那时的夕阳

正与此刻的晨曦

进行金属疲劳测试

The Polyphony of Steel

Twelve-meter beams slice through the dome of the sky,
Rivets sprout copper-green branches along the fault lines of
Memory.
Within the red brick walls,
The steam hammer of forty years ago still beats the rhythm of
Light.

Gears hover in mid-air,
Transforming into the voice of the bronze chime bells.
Hardened iron shavings are recrystallizing,
While the youth measure the stretch of steel with their
Scripts,
Through cadence and rhyme scheme,
Lines reflect sixty-four spectrums on the glass dome.

Cascading staircases spiral up toward the chimneys,
With each step embedding a history written in welds.
As moonlight bathes the steel,
Concrete columns crack,
Revealing veins of wisteria.
The cast iron beams,
Weathered by time,
Tremble softly as the camera's shutter clicks to life.

At midnight,
When the theater opens,
The steel skeleton begins to secrete poetry,
While abandoned hooks reach out for velvet curtains.
It's not rust oozing from the welded seams;
It's the youth that refuses to oxidize.

In the empty theater,

Brass valves still gulp in the tides of Time.

Every bolt hole serves as a porthole,

Offering glimpses into the hardened Dusk of years past,

The setting sun at that time,

Is conducting a metal fatigue test,

With the morning glow of this moment.

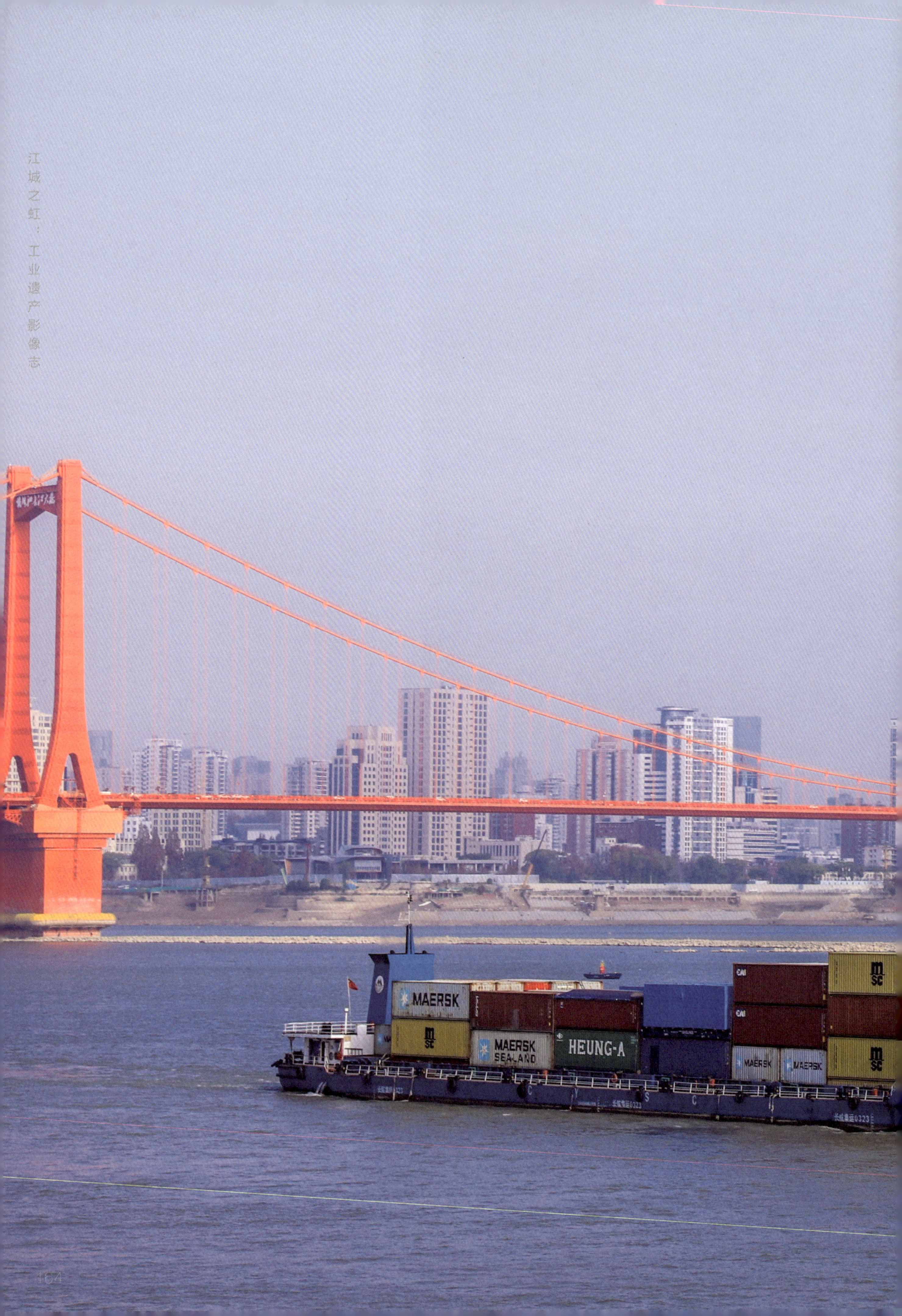
MAERSK
msc
MAERSK SEALAND
HEUNG-A
MAERSK
MAERSK

杨泗港都市 T 台

杨泗港都市 T 台位于武汉市汉阳区，北起汉阳江滩公园，南至杨泗港大桥江滩公园，全长约 2 千米。其前身是始建于 1959 年的杨泗港货运码头。杨泗港货运码头曾是长江中上游最大的国际集装箱转运码头。如今，这里已变身为集生态、文化、休闲于一体的滨江公园。杨泗港都市 T 台以"浮江展厅"为设计理念，保留了龙门吊、集装箱、铁轨等工业遗迹。其中，一座高 14 米、宽 20 米的橙色龙门吊被改造成水幕电影的"放映机"，夜幕降临时，水雾和光影交织，呈现出梦幻灵动的视觉效果。此外，杨泗港都市 T 台还设有玻璃栈桥、国风篮球场等特色设施。这里不仅是一个生态公园，更是武汉市民休闲娱乐的新晋热门"打卡地"。

Yangsigang Urban T-Stage

Yangsigang Urban T-Stage, located in Hanyang District, Wuhan City, stretches approximately 2 kilometers from Hanyang River Beach Park in the north to the Yangsigang Bridge River Beach Park in the south. Its predecessor was the Yangsigang Freight Terminal, originally built in 1959. Once the largest international container transshipment terminal in the middle and upper reaches of the Yangtze River, it has now been transformed into a riverside park integrating ecology, cultural heritage, and recreation. Designed with the concept of a "Floating River Exhibition Hall," Yangsigang Urban T-Stage preserves industrial relics such as gantry cranes, shipping containers, and railroad tracks. Notably, a towering 14-meter-high, 20-meter-wide orange gantry crane has been ingeniously repurposed into a water screen film projection installation. As night falls, mist and projected light intertwine, creating dreamlike and dynamic visual effects. Additional distinctive facilities include a glass walkway and a Chinese-opera-themed basketball court. Beyond being an ecological park, it has rapidly emerged as a popular new hotspot for leisure and entertainment among Wuhan residents.

江岸线解构练习

锈蚀物被集装箱抖落
在十二月的江面折叠成魔方
龙门吊将钢铁颈椎弯成满弓
将月光射向长江水幕

铁轨的断骨在沥青下苏醒
枕木的年轮被霓虹重新装订
当潮水漫过十四米橙色咽喉
集装箱豁口便吐出几十年前的汽笛
化作光粒子在薄雾中跳起探戈

玻璃栈桥剖开江水的琴键
浪花在环氧涂层上练习乐谱
篮球场将抛物线还给飞鸟
每次扣篮都惊起成群的集装箱
摆出莫比乌斯环的形态
在防波堤上舞蹈

高桩平台托起黄昏的天平
钢索吊起星空整片
为水幕电影更换菲林
而集装箱的棱角正被月光
锻打成新式编钟

子夜货轮拉响汽笛
工业骨骼扰动夜色
在潮汐退却前
献上最后一次洄游

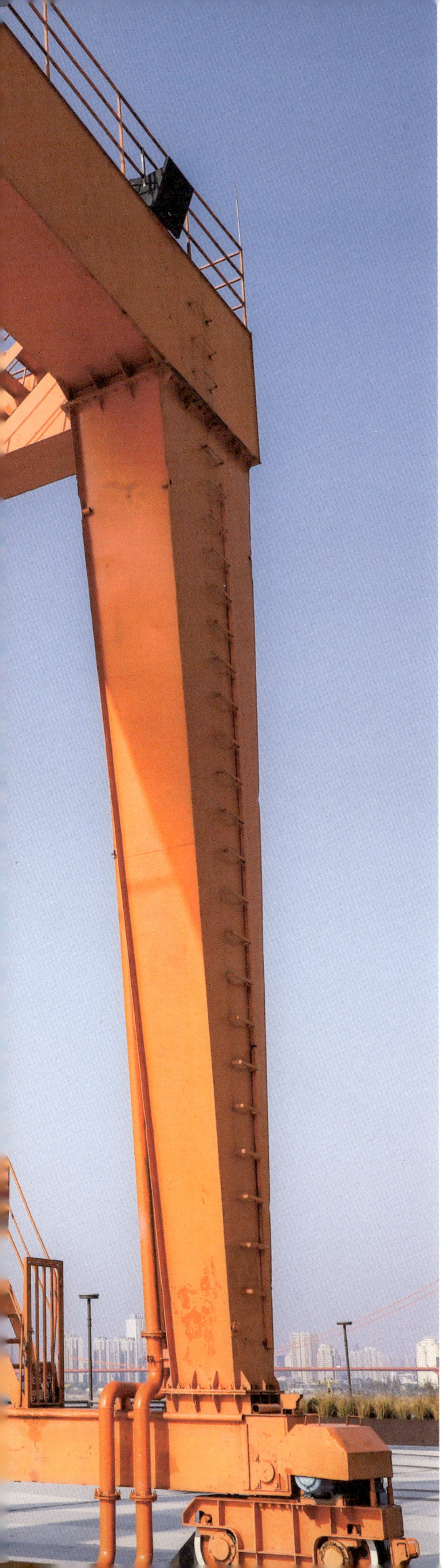

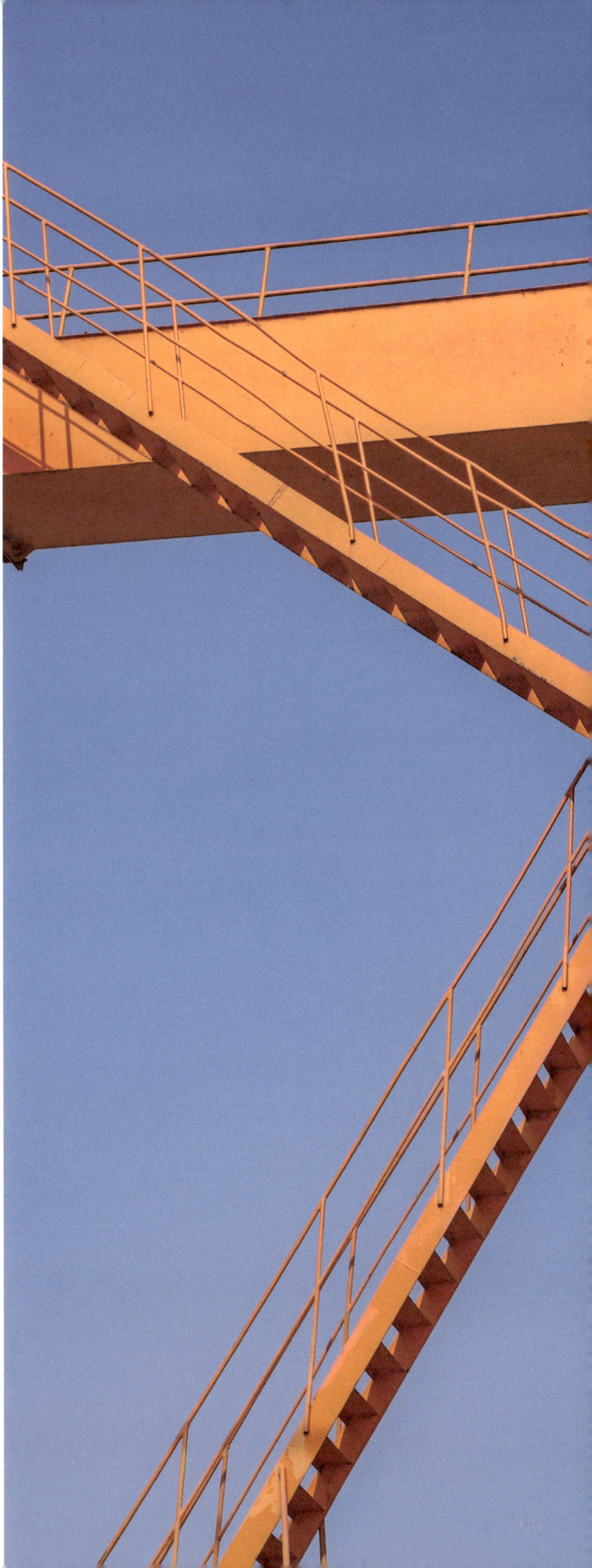

Exercise in Deconstructing the River Shoreline

Rusted remnants shaken free from containers,

Folded into a Rubik's cube on the river in December.

The gantry crane bends its steel cervical vertebrae,

A full bow that captures the whole Yangtze River,

Transforming it into a shimmering water screen.

The broken bones of the railroad track stir beneath the asphalt,

As decades of sleepers rebound with neon glow.

When the tidal wave rolls past the fourteen-meter orange throat,

Container openings exhale the sirens of several decades ago,

Turning into luminous particles swirling in a tango of mist.

The glass walkway dissects the river's musical score,

Waves practice a pentatonic melody on the epoxy surface.

The basketball court arcs back the parabola to the birds,

Every dunk startles the crowds of containers,

Forming the shape of a Mobius strip dance on the breakwater.

High-pile platforms cradle the scales of dusk.

The gantry crane hoists the entire starry sky with steel cables,

Ready to change the film for the water screen spectacle,

While the corners of the containers,

Kissed by moonlight,

Forge a new type of chime.

At midnight,

Cargo ships sound their whistles,

And every deconstructed industrial skeleton begins to swim.

Before the tide goes out,

Present the final migration of the steel structure.

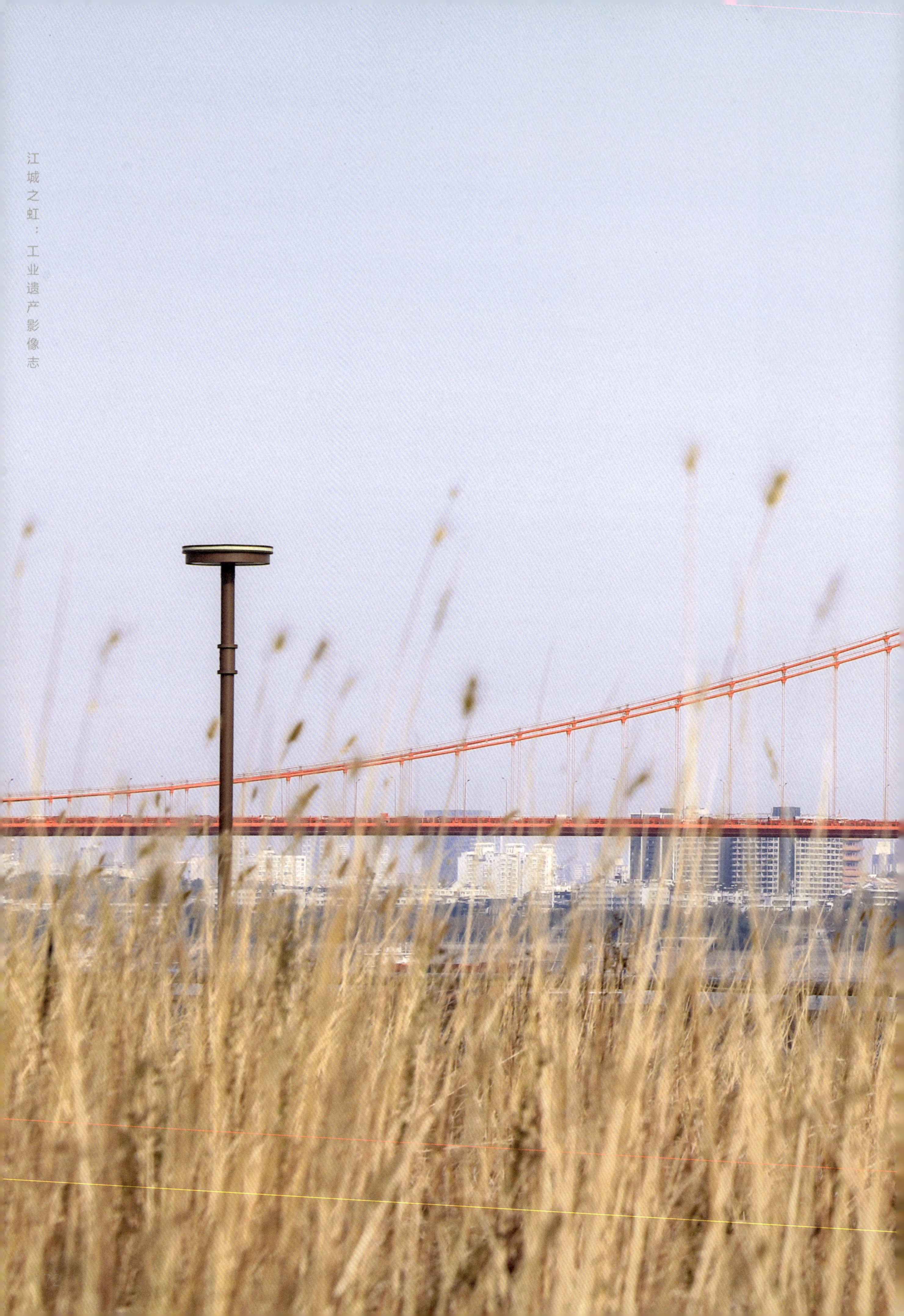

汉阳造文化创意产业园

汉阳造文化创意产业园位于汉阳区。这里曾是武汉鹦鹉磁带厂，企业改制后，厂区逐渐沉寂。后来，通过政府引导和市场运作，老厂房被重新定义和改造，成为集文化艺术、创意设计、商务休闲于一体的综合性文化创意产业园。园区东邻长江，西靠月湖，北为汉江，南枕龟山，周边有晴川阁、古琴台等景点，环境清幽，是武汉乃至全国较早成功改造的工业遗产文化创意园区。

"Hanyang Zao" Cultural and Creative Industrial Park

"Hanyang Zao" Cultural and Creative Industrial Park is located in Hanyang District. Once known as the Wuhan Parrot Magnetic Tape Factory, the site gradually fell silent after the enterprise restructuring. Through government guidance and market initiatives, the old factory was redefined and remodeled into a comprehensive cultural and creative industrial park that integrates culture and art, creative design, business, and leisure. With the Yangtze River to the east, Moon Lake to the west, Han River to the north, and Guishan to the south, the park is surrounded by attractions such as Qingchuan Pavilion and Guqin Terrace. Its tranquil environment makes it one of the earliest and most successful transformations of industrial heritage into a cultural and creative park in Wuhan and across China.

這有詩

磁轨上的解冻期

十二月的光线正擦拭着
铸铁窗棂的冰裂纹
磁带厂仓库张开豁齿的穹顶
放行十万粒悬浮的二氧化硅晶体

防震弹簧在玻璃地板下
蜷成青铜色问号
阳光以磁头的姿态切入
划开废弃的磁带山

三十年前的声波突然直立行走
穿过现代艺术装置的金属腔体
破碎的窗台结出冰凌的唱片
旋转时析出九十年代的车间民谣
院落的树影被纺织成黑胶纹路
而新栽的鹅掌楸正用枝丫
把寂静刻录成透明母带

那扇钉满铆钉的防火门后边
褪色的生产进度表正在融化
滴落成展厅的导览虚线
穿堂风
掀起了帘幕

我在哪里，都很想你
一如掀开流水线的混凝土槽
也如冬眠的晶体管忽然导通
或是被铝箔封存的啸叫与情书
正随冰晶在斜射光中
缓慢析出未完成的五线谱

此刻所有的铁锈都在反光
像卡带仓里等待消磁的往事
而解冻的聚酯薄膜正将此刻
压制成一首透明的 B 面单曲

The Thawing Period on the Magnetic Track

December light wipes ice cracks on cast iron
Window panes.

The open-toothed dome of the tape factory's
Warehouse releases a hundred thousand
Suspended silica crystals.

Shock-absorbing springs beneath the glass
Floor curled into bronze question marks,

As sunlight slices in like a magnetic head,

Cutting through an abandoned tape mountain.

Thirty-year-old sound waves suddenly stand
Upright,

Traversing the metallic bowels of modern art
Installations.

Broken window sills cradle ice-covered records,

Spinning and precipitating the workshop ballads
Of the nineties.

The shadows of trees in the courtyard intertwine
Into a vinyl pattern,

While the newly planted tulip trees use their
Branches to burn the silence onto a transparent
Master tape.

Behind the riveted fire door,

Faded production schedules are melting,

Dribbling into the dotted lines of the gallery's
Guide.

The wind in the hall lifts the curtain:

I miss you wherever I am,

Lifting the concrete chute of an assembly
Line,

Or in the sudden conduction of a hibernating
Transistor.

Or in the whistles and love letters sealed in
Aluminum foil,

As they slowly precipitate the ice crystals in
The slanting light,

Slowly interpreting the unfinished musical
Notation.

All the rust reflects now,

Like the past waiting to be demagnetized in a
Cassette bin.

And the thawed polyester film presses this
Moment into a transparent B−side single.

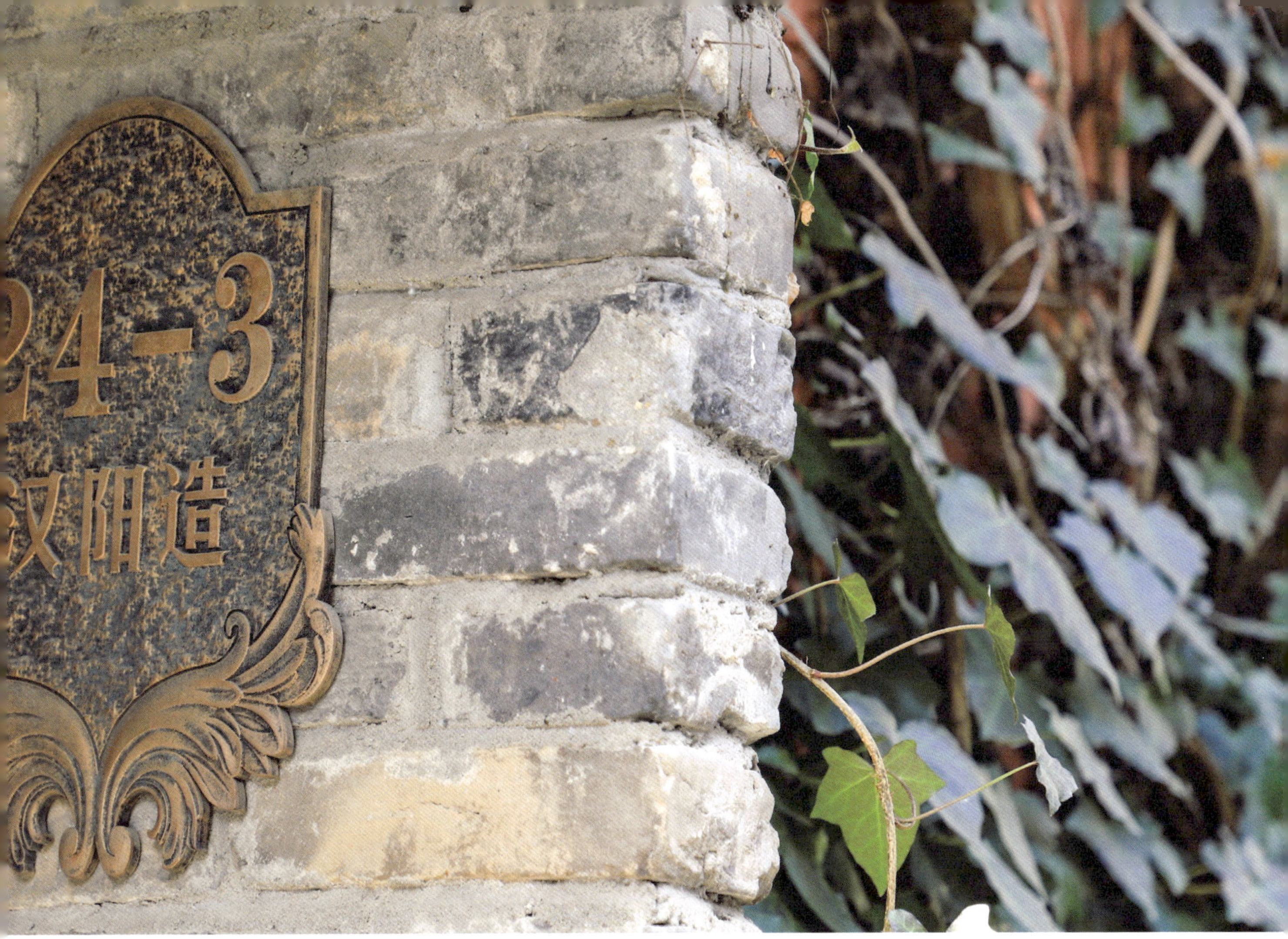
24-3
汉阳造

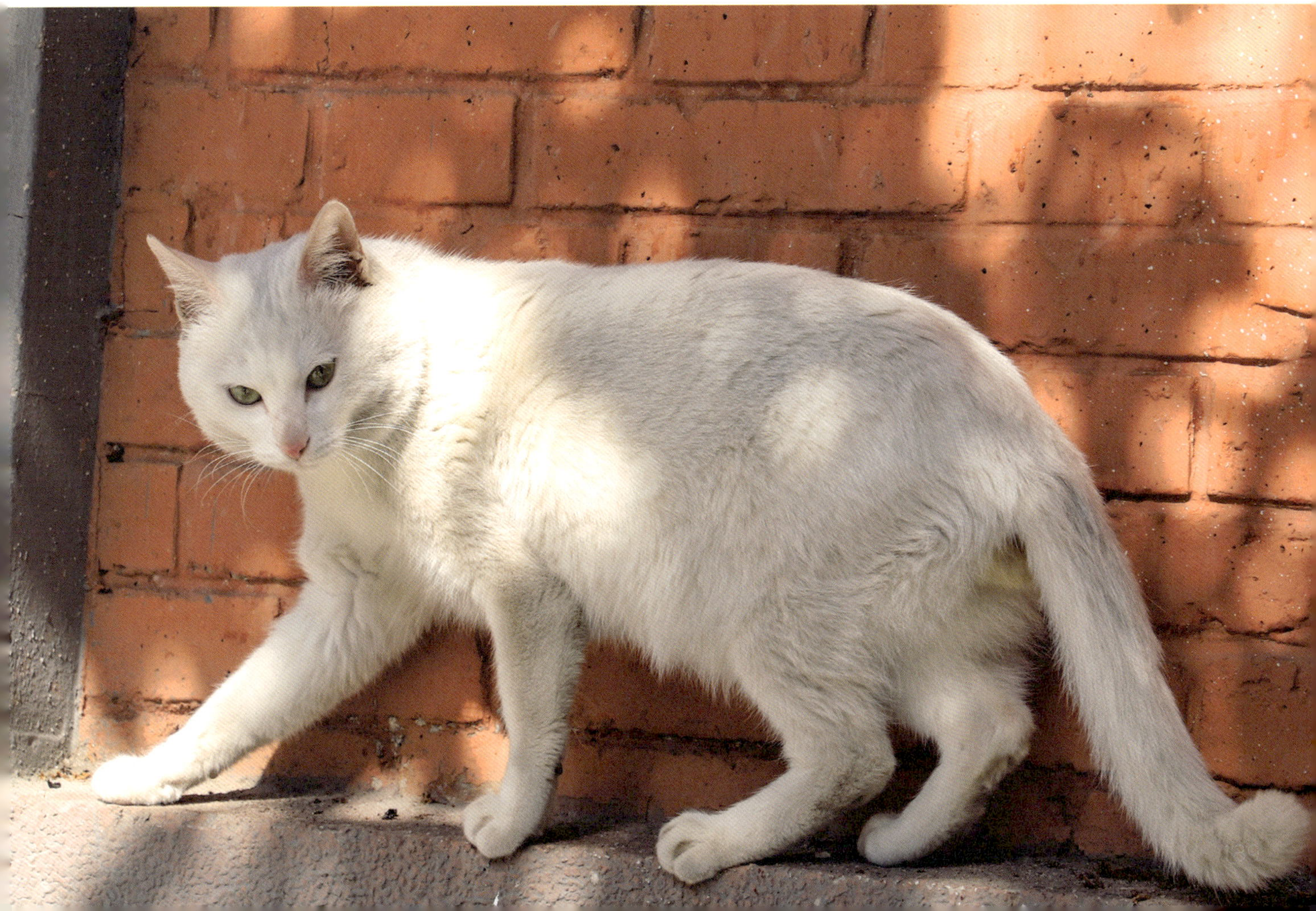

龙灵山生态公园

龙灵山生态公园位于武汉市蔡甸区，是以"生态修复"为主题的城市郊野公园，总面积约 290 公顷。公园依托龙灵山原有自然山体，通过生态修复技术，将昔日矿坑转化为绿意盎然的生态空间，成为武汉城市生态转型的典范。园内植被覆盖率较高，拥有十里花坡、九曲湾湿地等特色景观。春季芝樱盛开如粉色绒毯，秋季层林尽染，四季景致各异。全长 13 千米的环山绿道贯穿全园，途中设有多个观景平台及生态科普长廊，是徒步、骑行爱好者的理想之地。核心区域的九曲湾湿地栖息着 60 余种鸟类，形成独特的生物群落。龙灵山生态公园以矿坑变绿洲的奇迹诠释了生态治理的智慧，成为武汉生态文明建设的鲜活名片。

Longling Mountain Ecological Park

Located in Wuhan's Caidian District, Longling Mountain Ecological Park is an urban country park themed around ecological restoration. Spanning approximately 290 hectares, the park leverages the natural terrain of Longling Mountain, transforming former mining pits into a lush ecological haven through advanced restoration techniques. It stands as a model of Wuhan's urban ecological transformation. With high vegetation coverage, the park features signature landscapes such as Ten-Mile Flower Slope and Jiuqu Bay Wetland. In spring, Shibazakura blooms in a breathtaking pink cascade, while autumn paints the forests in rich hues, offering stunning scenery throughout the year. A 13-kilometer greenway winds through the park, featuring multiple viewing platforms and ecological science corridors, making it an ideal destination for hikers and cyclists. At the heart of the park, Jiuqu Bay Wetland is home to over 60 bird species, forming a unique and vibrant ecosystem. Longling Mountain Ecological Park is a testament to the power of ecological restoration—transforming a barren mine pit into a thriving oasis. It has become a shining example of Wuhan's commitment to ecological civilization.

绿焰

矿坑被时间炼成翡翠
山脊的裂痕翻涌成松涛阵阵
推土机锈蚀的齿缝间
芝樱花海抖开了粉色绸缎
将伤痕谱写成春的讯息

九曲湾的碧波里
白鹭叼起沉没的爆破声
涟漪在碎石上绣出年轮
沥青路蜿蜒
仿若青藤
缝补旧时裂纹

骑行道穿过银杏的指缝
安全帽在露水里酿制星光
碎石堆被鸟鸣重新上釉
每一粒泥土都发出新芽
向天空吐露翠绿的舌焰

蝴蝶羽翼在矿石间返青
山峰起伏在云边搬运绿意
松针戳破残存的雷管
当废墟的阴影淌成蜜时
春风恣意描摹满山的翠绿

The Green Flame

The quarry,

Tempered by time into jade,

Cracks along the ridge surge with pine waves.

In the rusted teeth of bulldozers,

Shibazakura seas shake out pink silks,

Composing spring's message from scars.

In Jiuqu Bay's emerald ripples,

Egrets pluck sunken explosions from the depths.

Ripples embroider rings of time on scattered stones.

Asphalt roads coil like ivy,

Stitching seams through cracks of yesterday.

Bike paths thread through ginkgo fingers,

Hard hats brew starlight in dewdrops.

Gravel heaps reglazed by birdsong—

Each grain of soil sprouts a green tongue,

Lifting jade flames toward the sky.

Butterfly wings regain verdancy among ores.

Mountains pulse,

Hauling green to the cloud's edge—

Pine needles pierce leftover detonators.

When shadows of ruins drip honey,

Spring wind sweeps the slopes wild with jade.

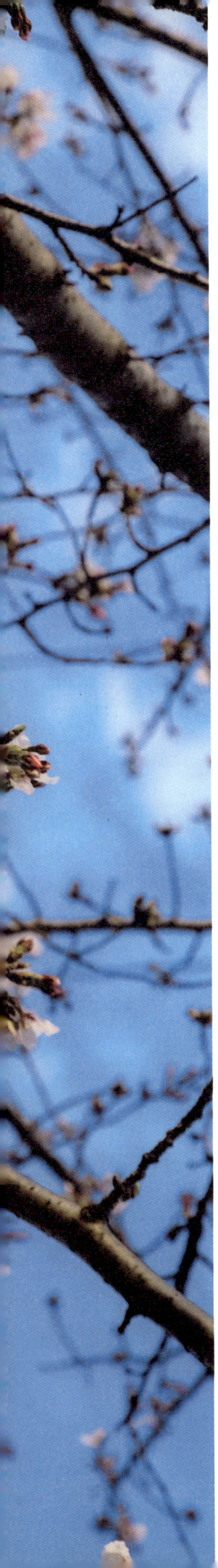
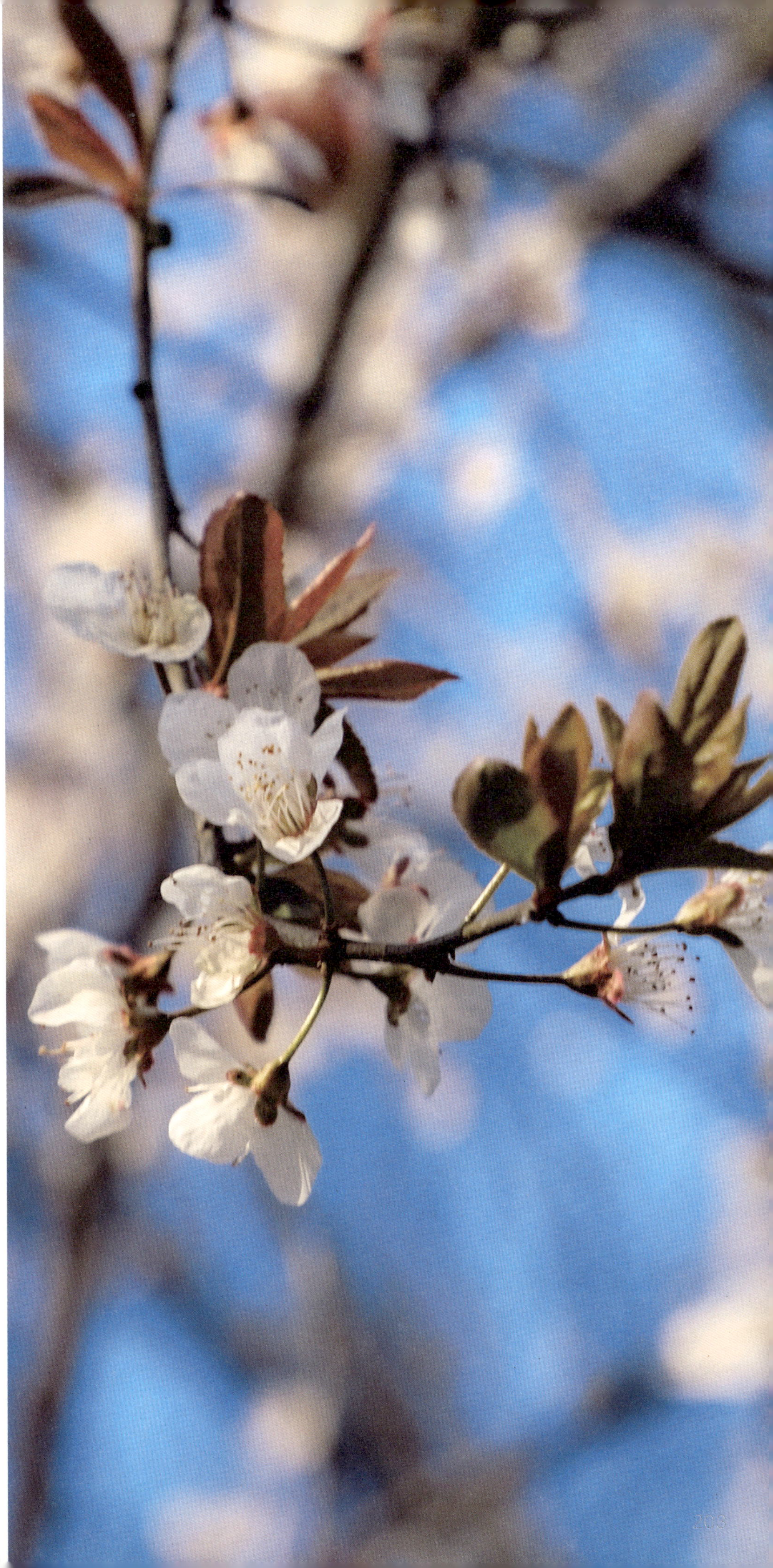

后记

江河的对话

按下快门的瞬间，我总听见，钢铁在呼吸。

在我一贯的认知里，"钢铁"二字，几乎与"冰冷"等同，因此，我也从不认为历经热熔锻造的钢筋骨架能与我产生任何联系。直至一个深秋，我走在粤汉铁路火车轮渡码头旧址边上，只见树影婆娑之下，江水浩浩汤汤，桁架斑驳不堪。江边的秋风瑟瑟而起，拂过水的裙角；江上的流水奔向岸边，浸湿风的衣袖。牵着风与水，我驻足望着耸立的钢架遗存良久。当时之思已难复原，只记得钢架上攀缘着层叠的贝壳，反复摩挲，粗砺之感久久未散。恍惚之间，我听见了百年前蒸汽机车的鸣笛声。

我与钢铁就此留在了那个深秋的早晨。

自那以后，我一次次站在由钢铁骨骼改造而成的穹顶之下，抬头注视着阳光射向残留的铆钉，在混凝土墙面投下时间的刻度。时间流逝代表着不断失去吗？我曾坚定不移地相信，答案唯有"当然"，但偏偏，我看见了被岁月锈蚀的承重梁仍在支撑着城市天际线，生锈齿轮咬合处萌发着的蒲公英依旧迎风起舞，龙门吊划定的抛物线所指向的始终是万千星辰——一座城市的壮美，恰恰在于它敢于将结痂的伤痕锻造成甲胄。失去的未必失去，正如被时间带走的泥沙，会在遥远的入海口铺成来时路；留下的始终留下，如同江水始终粼粼流淌，钢架挺立一如最初。

站在汉口江滩的吊机残骸前，我的取景框里同时盛着两个故乡。姚江的粼粼波光与长江的浑厚涛声，终于在我的镜头里跨越千里，拥抱交融。我知道，举起相机时心底的踏实感，正是源自江河的浪潮声。

作为一名土生土长的余姚人，我常常思忖，7000 多年前，先民们在姚江边抟土制陶时，是否也像我伫立在青山区红房子前这般屏息？那些用骨耜翻动沃土的手，与 20世纪 50 年代建设者抡起铁锤的手，是否拥有相似的指纹？宁波商帮的桅杆曾刺破的东海迷雾，与武汉锅炉厂的烟囱在 20 世纪 60 年代划开的天际，是否有着同一种开拓的魄力？

武汉的工业遗存，或傍水，或依山，即使点缀在城区，也多多少少保留着自然的气息。初冬时节，江边沿着铸铁蔓延的水汽，总让我想起每年清明，四明山上被云雾氤氲的香榧林。工业齿轮的咬合声早已消散，但透过取景器，我仍能看见昔日迸溅的火星，正与余姚通济桥的石缝里商贾的足迹遥相呼应。那些被江风侵蚀的铆钉，多像姚江古桥上的每一块石砖，静谧无言，却道尽了家长里短、日来月往。

感谢武大樱顶的月光，教会了我用新闻人的理性和诗人的感性双重曝光。遥望江水由西向东，我也终于明白，所谓工业遗产，不正是人类写给时间的情书，而长江与姚江，又何尝不是这些信笺最忠实的邮差。

这部摄影集是我的双城记，更是一封来自新时代的回信。不可否认，这是一个坐上了加速器的时代，慢慢地走，慢慢地看，慢慢地读，说是一种奢望都不为过。何况在大多数情况下，气喘吁吁地追赶人潮实为身不由己的选择。我也不例外。但我会努力拽住自己，时不时回到山里看一看，回到江边走一走，因为我知道，淬炼的火星也许会停歇，但河姆渡文化陶器上的稻穗纹永远金黄；亮起的手机屏幕也许会黯淡，但汉口码头的航标灯将照亮一代又一代人。

请允许我向最敬重的乡贤、散文巨擘余秋雨老师献上我最诚恳的谢意，感谢先生在百忙之中为拙作赐题书名。同样诚挚的感谢献给为拙作写序的武汉文史大家王汗吾先生，以及推荐拙作的肖佳法先生与向在荣教授。各位前辈对我的无私勉励，将支撑我义无反顾地继续前行。

此外，我还要向华中科技大学出版社金紫老师、责任编辑梁任老师与余秋雨先生的秘书金克林先生为拙作付出的心血由衷致谢，并特别感谢我在武汉大学的学术导师韩晗老师，韩老师对工业遗产研究的坚守始终激励着我。

最后，这本书献给我的父母，是他们对我的爱与支持，让我得以幸运地成长为一个勇敢而又充满热情的人。

胡晨冉
于珞珈山桂园